# SpringerBriefs in Electrical and Computer Engineering

**Series editors:**

Woon-Seng Gan, School of Electrical and Electronic Engineering
Nanyang Technological University, Singapore, Singapore
C.-C. Jay Kuo, University of Southern California, Los Angeles, CA, USA
Thomas Fang Zheng, Research Institute of Information Technology
Tsinghua University, Beijing, China
Mauro Barni, Department of Information Engineering and Mathematics
University of Siena, Siena, Italy

SpringerBriefs present concise summaries of cutting-edge research and practical applications across a wide spectrum of fields. Featuring compact volumes of 50 to 125 pages, the series covers a range of content from professional to academic. Typical topics might include: timely report of state-of-the art analytical techniques, a bridge between new research results, as published in journal articles, and a contextual literature review, a snapshot of a hot or emerging topic, an in-depth case study or clinical example and a presentation of core concepts that students must understand in order to make independent contributions.

More information about this series at http://www.springer.com/series/10059

Naghmeh Niknejad • Ab Razak Che Hussin
Iraj Sadegh Amiri

# The Impact of Service Oriented Architecture Adoption on Organizations

Naghmeh Niknejad
University Technology Malaysia (UTM)
Johor Bahru, Malaysia

Ab Razak Che Hussin
University Technology Malaysia (UTM)
Johor Bahru, Malaysia

Iraj Sadegh Amiri
Computational Optics Research Group
Advanced Institute of Materials
Science
Ton Duc Thang University
Ho Chi Minh City, Vietnam

Faculty of Applied Sciences
Ton Duc Thang University
Ho Chi Minh City, Vietnam

ISSN 2191-8112          ISSN 2191-8120   (electronic)
SpringerBriefs in Electrical and Computer Engineering
ISBN 978-3-030-12099-3        ISBN 978-3-030-12100-6   (eBook)
https://doi.org/10.1007/978-3-030-12100-6

Library of Congress Control Number: 2019930652

This Springer imprint is published by the registered company Springer Nature Switzerland AG
The registered company address is: Gewerbestrasse 11, 6330 Cham, Switzerland

# Preface

Service-oriented architecture (SOA) is one of the technologies in the area of information systems' design and architecture in the recent technology world. SOA adoption is known as an evolutionary process, instead of revolutionary one. In this process, an application is developed in a long period of time which will be improved gradually. Since there is a limited number of researches that focus on the effects of SOA adoption in an organization, this study investigated the significant factors which affected SOA and have been more discussible in the last 5 years based on previous studies such as governance, strategy, complexity, return on investment (ROI), business and IT alignment, culture and communication, costs, and security. Since most of the researches focus on qualitative analysis for SOA adoption, a need for empirical research was felt. So, this study conducted a quantitative analysis to investigate the influential factors for adopting SOA. In addition, an SOA adoption framework was proposed to measure the effect of factors on SOA adoption and the performance of organizations. Based on the proposed framework, an online questionnaire was created and distributed among SOA experts through LinkedIn (the largest social networking website for people in professional occupations) to collect data about the influence of SOA adoption on organizations. Subsequently, this study has made recommendations for improving the organizations planning to adopt SOA or in the way of adopting SOA to promote the performance of their organizations. On the other hand, the outcomes of this study may pave the way to form the basic knowledge in the domain of organizational and technological SOA adoption and trigger further research in the field.

Johor Bahru, Malaysia                                          Naghmeh Niknejad
Johor Bahru, Malaysia                                        Ab Razak Che Hussin
Ho Chi Minh City, Vietnam                                      Iraj Sadegh Amiri

# Abbreviations

| | |
|---|---|
| AVE | Average variance extracted |
| CIO | Chief information officer |
| CISO | Chief information security officer |
| CTO | Chief technology officer |
| IS | Information system |
| IT | Information technology |
| ROI | Return on investment |
| SmartPLS | Smart partial least squares |
| SOA | Service-oriented architecture |
| TOE | Technology-organization-environment |
| UTM | University of Technology Malaysia |
| VP | Vice president |

# Contents

# About the Authors

**Naghmeh Niknejad** received her B.Sc. from Islamic Azad University of Sepidan, Iran, in 2007. She got his M.S. in IT Management at the University of Technology Malaysia in 2014. She is a Ph.D. candidate in Information Systems at the University of Technology Malaysia. She has been doing research on several topics such as consumer behavior, technology management, information systems management (technology adoption), Internet of Things adoption, wearables adoption, and service-oriented architecture adoption.

**Ab Razak Che Hussin** is an Associate Professor of Information Systems at the University Technology Malaysia. He was awarded a Master's degree in Information Systems at the University of Technology Malaysia in 1998. He received his Ph.D. at the University of Manchester, UK, in 2006. His research focus includes studying information systems, web application, trust and privacy in e-commerce, database, IT planning and management, IT for organization transformation, and research methodology.

**Iraj Sadegh Amiri** received his B.Sc. (Applied Physics) from the Public University of Oroumiyeh, Iran, in 2001 and is a gold medalist M.Sc. (Physics/Optics) from the University of Technology Malaysia (UTM) in 2009. He was awarded a Ph.D. degree in Physics (Photonics) in January 2014. He has been doing research on several topics such as the optical soliton communications, laser physics, fiber lasers, fiber grating, electro-optical modulators, nanofabrications, semiconductor design and modeling, Lumerical modeling, plasmonics photonics devices, nonlinear fiber optics, optoelectronics devices using 2D materials, semiconductor waveguide design and fabrications, photolithography fabrications, e-beam lithography, quantum cryptography, and nanotechnology engineering.

# Chapter 1
# Introduction of Service-Oriented Architecture (SOA) Adoption

## 1.1 Overview

SOA adoption is known as an evolutionary, instead of revolutionary, process. In other words, adopting SOA is similar to a trip for an organization in a long period of time. It is not developing an application in a short period of time [1]. Specialists in many fields are concerned with organizational performance including strategic planners, operations, finance, legal, and organizational development [2].

SOA seems different for different users. In other words, it is different from various perspectives. For example according to [3], SOA has an architectural form and needs criteria, patterns, architectural principles, service description, requestor, and a service provider that shows attributes like encapsulation, modularity, segregation of concerns, loose coupling, reuse, and so forth. SOA is a model of programming that is completed with standards, technologies like web services, and a solution for middleware which optimize for monitoring, orchestration, service assembly, and a management.

While more firms around the world have started to search about SOA, some implementation topics are represented that some aspects of implementation are undervalued like complexity, cost, and the attempt necessary for even a little enhancing to implement SOA [4]. A report expressed SOA impact organizations positively and negatively at the same time. When SOA positively influences on modifiability, extensibility, and interoperability, on the other side it affects performance, auditability, testability, and security negatively [5]. So, the main aim of this study is to evaluate the impact of SOA adoption on organizations.

© The Author(s), under exclusive license to Springer Nature Switzerland AG 2019
N. Niknejad et al., *The Impact of Service Oriented Architecture Adoption on Organizations*, SpringerBriefs in Electrical and Computer Engineering,
https://doi.org/10.1007/978-3-030-12100-6_1

## 1.2   Background of Study

Growing competitiveness, globalization, and ever faster creativity are the specification of modern economies. It is highly worth mentioning that a developing move towards new markets, a responsive change towards business strategies, or providing effective feedback to competitive pressures pave the way for the organizations to focus on a high level of flexibility [6].

Currently, the preferred architectural design to supply organizational agility, to promote application adaptability and system interoperability, and to provide the reuse of legacy possessions referred to service-oriented architecture, henceforth SOA [4]. Consecutive innovation, competitive emprise, and agility are changing into an important component of strategic thinking in a large number of current organizations. As a result, the growing of information systems has given birth to many organizations to re-evaluate their techniques as well as re-examine information technology function in forming their business strategies [7].

It is astonishing that although service-oriented architecture has been used for a decade, only several research studies have been performed on critical aspects that must be concentrated on during such implementations. Results show that there are a number of resemblances to success factors found in attaining strategic alignment, such as top management support and communication between collaborating parties [8].

SOA is formed on the base that systems are divided into sub-systems—each managing individual tasks—based on group responsibility in the business course of a company and then eventually all the responsibilities are seen as an interoperable service [1]. So, considering new information system, SOA is a complex solution of analysis, design, conserving, and combination of enterprise application dependent on services. Services are considered as unconnected program-autonomous existence which supplies one or more operational capabilities through their interface [9].

The fleeting environment of business organization challenges the flexibility and adaptability capabilities of organizations. It can be claimed that every IT manager is looking into SOA [10]. SOA adoption is different from developing an application which is done in a short period of time. Therefore in applying SOA in organizations many problems appear, e.g., immature standards and inadequate knowledge [11].

Naturally, as other different kinds of technologies, some groups of people accept SOA as a perfect and precise technology and this is when other groups of people reject it for being imperfect. But, it is crystal clear that no one tends to ignore the achievements which SOA has brought about in the cases of efficiency, reusability, agility, and productivity of an enterprise [12].

As identified in various sources, SOA is a developed method in the direction of IT and information system architecture based on a cluster of services which are in relation with each other. A highly business-special definition introduces service-oriented architecture as a design that applies roles via reusable business services [13].

Not only IT but also business perspectives pave the way for the most thorough definition of Service Oriented Architecture. SOA provides an architectural model that tends to promote the efficiency, practicality, agility, and productivity of an enterprise by determining services as the main tool through which solution logic is demonstrated in support of the understanding of the strategic goals related to service-oriented computing [12]. In order to make a great progress, organizations are supposed to modify their approaches, types of communication, channels of collaborating, and techniques of publishing relationships, so these are the most challenging difficulties of organization and governance [14].

A famous technology and market research company, Forrester Research, executed a study claiming that the uses of SOA have been developed from 44% in North-American, European, and Asian-Pacific companies to 63% [15]. On the other hand, different conferences on the pattern of SOA were held since 2002 such as International Conference on Web Services (ICSOC), the IEEE International Conference on Web Services (ICWS), the International Conference on Services Computing (SCC), and the European Conference on Web Services (ECOWS). Consequently, a substantial SOA awareness exists in the domain of educational studies and industry practitioners and it is assumed that many companies performing something connected to SOA [16].

Providing the chance of fee-based services is an outstanding value that SOA serves to various companies of different sizes from small to medium-sized ones and it is not just in the possession of big organizations [6]. In spite of that, the 2009 Forrester SOA research demonstrated that in smaller organizations SOA adoption is much inferior, namely the companies with less than 1000 staffs [17]. The desire of SOA adoption is not satisfactory as it supposed to be [18, 19]. Claims on the part of some industries proposed that SOA faced a failure at presenting its suggested and guaranteed benefits and it turns to be of a high expense [20].

## 1.3   Problem Statement

In spite of a large number of educational cases connected to SOA, it is being discussed that academic background and related literature is to some extent disintegrated and untimely considering why and to what extent companies accept SOA. Furthermore, there is a limited number of researches that focus on the effects of SOA adoption in an organization [21].

The principal question of how SOA can promote organization agility and nurture closer alignment between IT and business has not been appropriately focused on. The vigorous communication among external business environmental factors, organizational agility, and IS architecture lead to a high complexity of the process of keeping IT and business in alignment [22].

One frequent problem in SOA adoption is that many associations initiate the project of adopting SOA with regard to an IT prospect rather than a business one. Regarding the technical perspectives of the project, implementations might emerge successful, but the effect of the adoption of the new architecture on the business cannot be recognized without having been counted right from scratch. Difficulties like these are mainly noticed in large associations with well-founded IT departments which try to obey every new fashion of technology. Not amazingly, the lack of business alignment with the SOA movement project is a definite result of such infirm project planning. The most feasible rejecting consequence of such usual mistakes is the increasing expense of IT without any return on investment (ROI) for the corporation [23].

As far as the researcher knows, there has been little research done to examine the effect of the adoption of service-oriented architecture on the performance of organizations.

## 1.4 Research Questions

Based on the abovementioned problem, the present research attempts to answer the following questions:

1. Q1: What are the factors which influence SOA adoption in organization?
2. Q2: What are the relationships among significant factors, SOA adoption and the performance of organizations?
3. Q3: What are the recommendations for organizations towards success of SOA adoption?

## 1.5 Objectives of Study

In the light of the problem statement, the present study aims to determine the following objectives through which organizations can make progress:

1. To identify the factors influenced by adoption of SOA in organizations.
2. To understand the relationship among significant factors, SOA adoption and the performance of organizations.
3. To develop recommendation towards success of SOA adoption.

## 1.6 Scopes of Study

As it was mentioned before, the purpose of this study is to discover the factors that affect SOA adoption in an organization and estimating the impact of SOA adoption on the performance of enterprises. In order to narrow the scope of this study down and based on the objectives of this research, the researcher focuses on organizations which have already adopted SOA. Data will be collected from skilled experts in SOA all over the world using online questionnaire. Some technical and organizational concepts will be presented in this study, but its purpose is to present a general horizon of SOA adoption and its performance.

## 1.7   Significance of Study

A number of limited researches exist studying the function of SOA adoption in organizations. This study tends to show how SOA adoption may pave the way for the organizations to make progress, by giving them the opportunity of being in an active access to the SOA. The results of this study might help organizations which adopt SOA to promote their chance of development.

### 1.7.1   Theoretical Implication

Help organization to find out the key factors which influence SOA adoption to improve organizational performance.

Add performance of organization to the TOE framework would give this idea to organization to improve their performance by adopting SOA.

### 1.7.2   Practical Implication

Help organization to focus on significant factors to accelerate the process of SOA adoption.

Improve performance of organizations by successful adoption of SOA.

Develop recommendations based on SOA experts' perspectives for being successful in SOA adoption.

This study tends to show how SOA adoption may pave the way for the organizations to make progress, by giving them the opportunity of being in an active access to the SOA.

The results of this study might help organizations which adopt SOA to promote their chance of development.

The findings of this study will encourage those companies that have not been using SOA to improve their strategies by adopting SOA.

### 1.7.3   Methodological Implication

Use self-selected sampling and distribute online questionnaire among SOA professionals through LinkedIn.

## 1.8 Structure of Study

This study is consisting of six chapters as shown in the following Fig. 1.1:

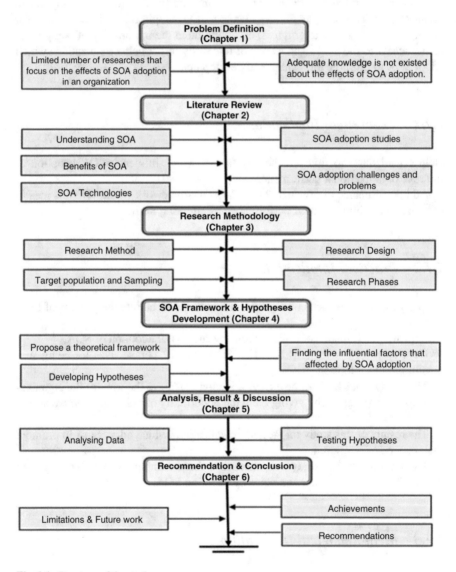

**Fig. 1.1** Structure of the study

## 1.9 Thesis Outline

In this chapter, the researcher focuses on introduction and background of the study about the key points of the study and problem background in terms of the clear background study of project. Moreover, research questions and research objectives have been discussed. Final steps consist of significance of the study, scope, structure, and outline of the project.

In Chap. 2 based on the problem of the study the academic literature will be studied. In this chapter, author focuses more on the SOA adoption in an organization and extracts significant factors influence by SOA adoption based on previous studies. Then TOE framework for adoption will be described.

In Chap. 3, the research methodology of the study includes the methods which are used in this research. The questionnaire will be distributed among the SOA experts and the results will be explained at the end of this chapter.

In Chap. 4, influential factors that concluded from papers reviewed in Chap. 2 will be investigated as the initial findings of this study. Then the theoretical framework and hypotheses will be proposed in continuation. Afterwards, pilot study will be performed. Final step includes the explanation of validity and reliability of the questionnaire.

In Chap. 5, the researcher will discuss the data collection and the data analyses. Data will be analyzed with SmartPLS. In continuation, the hypotheses of this study will be examined. The relationship among the significant factors, SOA adoption and the performance of organizations will be evaluated at the end of this chapter.

In Chap. 6, the achievements and the contribution of this study will be explained shortly. Then few recommendations will be discussed based on SOA experts' experiences. Moreover, some limitations and direction for future work will be described at the end of this study.

## References

1. T. Erl, *Service-Oriented Architecture Concept, Technology and Design* (Pearson Inc., Upper Saddle River, 2005)
2. S. Marisoosay, *The Role of Employee Turnover Between Employee Job Satisfaction and Company Performance in the Penang Automation Industry* (USM, 2009)
3. S. Chan, Smart SOA in Action-SOA Service Creation (IBM Corporation, 2009), [Online], ftp://public.dhe.ibm.com/software/hk/e-business/events/.../SOA_Service_Creation.pdf%0A. Accessed 12 Jun 2012
4. G.A. Lewis, E. Morris, S. Simanta, L. Wrage, Common misconceptions about service-oriented architecture, in *2007 Sixth International IEEE Conference on Commercial-off-the-Shelf (COTS)-Based Software Systems (ICCBSS'07)*, IEEE, 2007, pp. 123–130
5. L. O'Brien, P. Merson, L. Bass, Quality attributes for service-oriented architectures, in *Proceedings of the International Workshop on Systems Development in SOA Environments*, IEEE Computer Society, 2007, p. 3
6. D.K. Barry, *Web Services and Service-Oriented Architectures: The Savvy Manager's Guide* (Morgan Kaufmann, Burlington, 2003)

7. V. Sambamurthy, A. Bharadwaj, V. Grover, Shaping agility through digital options: reconceptualizing the role of information technology in contemporary firms. MIS Q. **27**, 237–263 (2003)
8. S. Emadi, R.H. Hanza, Critical factors in the effective of service-oriented architecture. Adv. Comput. Sci. **2**, 26–30 (2013)
9. P. Weiss, *Modeling of Service-Oriented Architecture: Integration of Business Process and Service Modeling*, vol 5 (STU Press, Vazovova, 2010), p. 07
10. M.P. Papazoglou, W.-J. Van Den Heuvel, Service-oriented design and development methodology. Int. J. Web Eng. Technol. **2**, 412–442 (2006)
11. S. Tilley, J. Gerdes, T. Hamilton, S. Huang, H. Müller, D. Smith, K. Wong, On the business value and technical challenges of adopting web services. J. Softw. Maint. Evol. Res. Pract. **16**, 31–50 (2004)
12. T. Erl, *Soa: Principles of Service Design* (Prentice Hall, Upper Saddle River, 2008)
13. M. Galinium, N. Shahbaz, Factors affecting success in migration of legacy systems to service-oriented architecture (SOA), Master Thesis, Department of Informatics, Lund University, Sweden, Submitted, 2009
14. R. Varadan, K. Channabasavaiah, S. Simpson, K. Holley, A. Allam, Increasing business flexibility and SOA adoption through effective SOA governance. IBM Syst. J. **47**, 473–488 (2008)
15. R. Heffner, SOA Adoption: Budgets Don't Matter Much (Forrester, 2008), [Online], https://www.forrester.com/report/SOA+Adoption+Budgets+Dont+Matter+Much/-/E-RES45277. Accessed 10 Jul 2011
16. H. Luthria, F. Rabhi, Using service oriented computing for competitive advantage, in *AMCIS 2009 Proceedings*, Paper 140, 2009
17. J. McKendrick, Forrester: only one percent have negative experience with SOA (ZDNet, 2009), [Online], https://www.zdnet.com/article/forrester-only-one-percent-havenegative-experience-with-soa/. Accessed 24 Jun 2010
18. K. Kontogiannis, G. Lewis, D. Smith, The landscape of service-oriented systems: a research perspective for maintenance and reengineering, in *Proceedings of the Workshop on Service-Oriented Architecture Maintenance*, Amsterdam, 2007
19. M.N. Haines, The impact of service-oriented application development on software development methodology, in *2007 40th Annual Hawaii International Conference on System Sciences (HICSS 2007)*, IEEE, 2007, pp. 172b–172b
20. M. Meehan, SOA adoption marked by broad failure and wild success (TechTarget, 2008), [Online], https://searchmicroservices.techtarget.com/news/1319609/SOA-adoption-marked-by-broad-failure-and-wildsuccess. Accessed 20 May 2012
21. N. Joachim, D. Beimborn, P. Hoberg, F. Schlosser, Examining the Organizational decision to adopt service-oriented architecture (SOA)—development of a research model. Digit 2009 Proc. Diffus. Interes. Gr. Inf. Technol. (2009)
22. J. Choi, D.L. Nazareth, H.K. Jain, The impact of SOA implementation on IT-business alignment: a system dynamics approach. ACM Trans. Manag. Inf. Syst. **4**, 3 (2013)
23. L. Cherbakov, M. Ibrahim, J. Ang, SOA antipatterns: the obstacles to the adoption and successful realization of service-oriented architecture (2006), [Online], www.ibm.com/developerworks/webservices/library/ws-antipatterns/. Accessed 1 Feb 2010

# Chapter 2
# Literature Review of Service-Oriented Architecture (SOA) Adoption Researches and the Related Significant Factors

## 2.1 Overview

Based on the reviewed papers, this chapter selected the potential factors from 2009 until 2013. At the end of this section, TOE framework is explained briefly and three SOA adoption models based on TOE framework are described. All the information and definitions are conducted according to previous work on SOA adoption through surfing the net, reading prior journals and papers. The main purpose of this chapter is to focus significant points about adoption of SOA in organizations. The following figure shows the whole structure of this section (Fig. 2.1).

## 2.2 Service-Oriented Architecture

### 2.2.1 Understanding SOA

Service-Oriented Architecture (SOA) is one of the most remarkable technologies in the area of information systems' architecture and design in the modern technology world. SOA is an approach to look at the world. With a service-oriented observation, everything seems like a service. The service is a fundamental construction unit of SOA. It is a technique of accessing repeatable business capabilities [1]. According to Erl [2], systems are separated into sub-systems based on group functionality in the business process of an organization. Each sub-system managing individual tasks and all the functionalities are grouped as an interoperable service finally. SOA is an approach to look at the world. With a service-oriented observation, everything seems like a service. The service is the fundamental unit of SOA. It is a technique of accessing repeatable business capabilities.

N. Niknejad et al., *The Impact of Service Oriented Architecture Adoption on Organizations*, SpringerBriefs in Electrical and Computer Engineering, https://doi.org/10.1007/978-3-030-12100-6_2

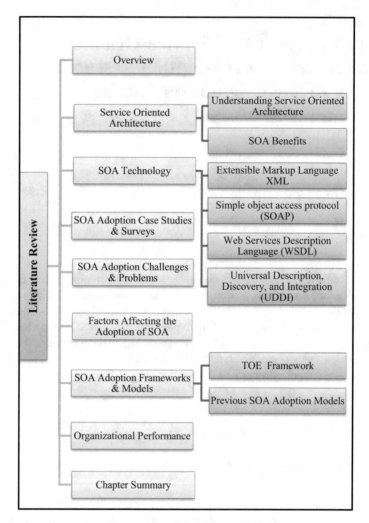

**Fig. 2.1** The structure of literature review

Service-oriented is used as a term of IT/IS for some time. It is used in various areas. Regardless of the diverse use of this term, there is a general viewpoint which represents a special view to split tasks as a solution to solve problems. It is a standard that helps the necessary reasoning to answer the problem. Service-oriented is splitting the problems into entity and related smaller parts of logic or service. Service orientation standard proposes the organization needs to re-explain businesses tasks, entity, or process into well-grained service granularity from the perspective of IT/IS. A well-grained service guarantees their individuality. It makes them to be easier to compose and orchestrate with all other services [3].

Systems are separated into sub-systems based on group functionality in the business process of an organization. Each sub-system controls independent tasks.

**Table 2.1**  Service-oriented architecture definitions

| No | Author/year | SOA definition |
|---|---|---|
| 1 | Natis (2003) Gartner Group | "…a software architecture that starts with an interface definition and builds the entire application topology as a topology of interfaces, interface implementations and interface calls. SOA would be better named interface oriented architecture. SOA is a relationship of services and service consumers, both software modules large enough to represent a complete business function. Services are software modules that are accessed by name via an interface typically in a request-reply mode. Service consumers are software that embeds a service interface proxy (the client representation of the interface)" |
| 2 | Lublinsky (2007) | "SOA can be defined as an architectural style promoting the concept of business-aligned enterprise service as the fundamental unit of designing, building, and composing enterprise business solutions" |
| 3 | Bieberstein, Bose et al. (2005) | "Framework for integrating business processes and supporting IT infrastructure as secure, standardized components—services—that can be reused and combined to address changing business priorities" |
| 4 | Krafzig, Banke et al. (2005) | "A Service-Oriented Architecture (SOA) is a software architecture that is based on the key concepts of an application frontend, service, service repository, and service bus" |
| 5 | Erl (2005) | "SOA initially is as an abstract paradigm represented a baseline for distributed architecture that has no reference for its implementation (traditional form), then it evolves as an architectural style that delivers service orientation through the use of web services (contemporary form)." And a formal definition: "SOA is a form of technology architecture that adheres the principles of services orientation" |
| 6 | McGovern, Sims et al. (2006) | "SOA as component based software modules that provide service to other modules" |

Finally, all the functionalities are grouped as an interoperable service. Actually there was a growing trend about SOA and its adoption within a variety of different scale enterprises in the previous decade. IT leaders anticipate a bigger raise in SOA adoption according to the promised benefits acquired by early adopters [2] (Table 2.1).

The abovementioned descriptions have diverse perspectives to realize SOA. For instance, Erl [2] initially perceived SOA as abstract model from distributed architecture viewpoints while Bieberstein et al. [4] view SOA from business and technical perspective, and McGovern et al. [5] view from component-based software perception.

### 2.2.2   SOA Benefits

According to Grigoriu [6], usually the most important benefits collaborated to SOA are the reuse of technology and agility. The frequently used approach of SOA out of an enterprise architecture context is advanced gradually, without interfering the benefits brought by the enterprise architecture. As the result, the agility is not gained or is gained late, until the end of the journey to migrate to SOA of the enterprise while the reuse of technology may need to redesign costly. In fact the reuse of business process is more significant than the reuse of IT, because SOA determines the same business activities and puts them in a group as a service. SOA will decrease the application duplication by reducing process replication [6].

According to Jayashetty and Kumar [7], SOA improves revenue via enhancing business agility by adopting renovated business models and proposing new products and services in the right time. SOA also reduces cost via separating the implementation features from the service consumers, increasing reusability, and eliminating unnecessary duplication in the system. In a study, Yoon and Carter [8] determined the precursor and advantages of selecting SOA. In this multiple case study, they emphasized on the influence of using SOA from the standpoint of business value. The research team divided the benefits of SOA adoption in two categories: business agility and cost.

Newcomer and Lomow [9] and Erl [2] provide the benefits of SOA in a detailed way. Newcomer and Lomow discuss about the difference between technical benefits and business benefits. All business benefits are covered by the classification of Erl excluding increased customer satisfaction. The technical benefits of Newcomer and Lomow can be interrelated to the "reduced IT burden" benefit from Erl [10].

According to Erl [2], the following are the benefits and attributes of services in a service-oriented architecture (Table 2.2):

## 2.3   SOA Technology

One of the most popular technologies to implement SOA is web services. A web service defines a standard way to integrate web-based applications by using technologies like Extensible Markup Language, Simple Object Access Protocol, Web Services Description Language and Universal Description, Discovery and Integration. Web services are independent. It means that their function is not related to any operating system or programming language [11].

A web service is a part of software that acts with set of standards reciprocally. These standards enable global interoperation of computers regardless of operating systems, hardware platform, programming language, or network infrastructure. Web services depend on three interrelated XML-based software standards to work; SOAP, WSDL, and UDDI. Basically web services exchange SOAP messages. Services using multiple interfaces and a description language (WSDL) can be called to perform business processes. Each interaction is free of other interaction. The aptitude of a web service to work anywhere on any network with no impact on its performance is depended on network transparence. Since every web service has special characteristics, web services include the same flexibility to

**Table 2.2**  The benefits and attributes of services in a service-oriented architecture [2]

| Service | Description |
| --- | --- |
| Reusable | Services are planned to support potential reuse even if immediate reuse opportunities exist |
| Stateless | Services are work independently without caring about the last or next tasks they asked to do |
| Discoverable | A service should be discoverable by service requesters. It means that when a consumer searching for a service it should be discovered and invoked from SOA service directory |
| Self-describing | The interface of services should explain, reveal, and present an "entry point" to the service. The interface of service should cover all the information about the service, so a service customer could find and connect to the service easily. It avoids requiring the consumer to recognize the technical implementation details |
| Composable | SOA services are composite intrinsically. They may be created from other services and be joined with other services to create other business solutions |
| Single-instance | In single instance, only one implementation of a specified service should be in an SOA |
| Loosely coupled | It allows application features to be alienated into independent parts. This "separation of concern" prepares a mechanism for services to contact others without firmly bounding to each other |
| Governed by policy | There are services that are built by contract. Interactions between services and between services and service domains are controlled by policies and service-level agreements, advancing process efficiency and decreasing complexity |
| Independent of location, language, and protocol | Services are intended to be available to any certified user from any location, on any platforms and with a general speaking |
| As coarse-grained as possible | Granularity is a declaration of functional wealth for a service. It means if a service is more coarse-grained, the function offered by the service is richer. This attitude decreases complexity for system developers by restricting the necessary steps to completing a given business function |
| Potentially asynchronous | Asynchronous improves system scalability during asynchronous performance and queuing techniques. High connections expenses and changeable network latency can deliberate response times in an SOA environment. Because of the disseminated nature of the services, asynchronous behavior allows a service to concern a service demand and continue processing to get a response by the service provider |

websites on the internet. It can be situated on any computer that is linked to the network by internet protocols. Web services are efficient techniques to practice service-oriented architecture, since the standards and infrastructures which support these technologies are finally available to make web services based on SOA practical [12, 13].

### 2.3.1  Extensible Markup Language (XML)

World Wide Web Consortium defined XML as a text-based markup language. Contrasting with HTML, which apply tags for describing performance and information, XML is for the description of portable ordered data. It is utilized as a language for describing data description languages, like markup grammars, vocabularies, interchange formats, and messaging protocols [14].

Also, XML technologies propone many other benefits: integration of organization data from different sources; the flexible connection between managed objects and organization application; interoperability between management applications from diverse vendors; simple and powerful description and transformation of information managing; routine and central validation of managing data. These benefits and other technologies, well matched with SOA architecture, offer a working solution to apply dynamic e-business solutions [14].

### 2.3.2  Simple Object Access Protocol (SOAP)

SOAP is an XML-based protocol for exchanging information in a decentralized, dispersed environment. SOAP is a messaging protocol that transfer structured information between applications or systems. Then the requesting objects are able to make a distant method request on the preparing objects in an object oriented programming manner. The simple object access protocol requirement is provided by User Land, Lotus, IBM, Microsoft, and Develop Mentor. The requirement spawned the foundation of the W3C XML Protocol Workgroup, contained more than 30 companies. The basis of dispersed object communication in vendor implementations of SOA is formed out by SOAP. However, SOA does not represent a messaging protocol. Since SOAP is commonly used for implementing SOA, it has been named as SOA Protocol. One of the advantages of SOAP is that it is totally vendor-neutral which allows implementing independently of the platform, object model, programming language and operating system. Moreover, data-encoding preferences, language bindings, and transport implementation are dependent [13–15].

### 2.3.3  Web Services Description Language (WSDL)

It is an XML word that prepares a standard technique of elaborating service IDLs. A convergence of activity between SDL (Microsoft) and NASSL [16] results in WSDL. It presents an easy way for service providers to explain the template of request and respond messages for remote method invocations (RMI). WSDL leads this title of service IDLs independent of the fundamental protocol and encoding necessities. An abstract language for explaining the operations of a service with data types and particular parameters is prepared by WSDL. The description of the setting and binding features of the service are addressed by the language as well [14, 15].

### 2.3.4 Universal Description, Discovery, and Integration (UDDI)

An ordinary team of SOAP APIs that empower the accomplishment of a service mediator is provided by the Universal Description, Discovery, and Integration. In order to pave the way of accelerating the description, discovery, and integration of web-based services, the UDDI determination was specified by IBM, Microsoft, and Ariba. It is important to understand the features of the SOA better to identify how many components of the SOA operate mutually. A service provider establishes a web service and its explanation and then publishes the web service in UDDI. When a web service is published, a service requester may discover the service using the UDDI connection. The UDDI registry prepares the service requester with a WSDL service explanation and a uniform resource locator (URL) referring to the service. Finally, the service requester invokes the service by using this data to attach it [14, 17]. The architecture of web service is shown in Fig. 2.2.

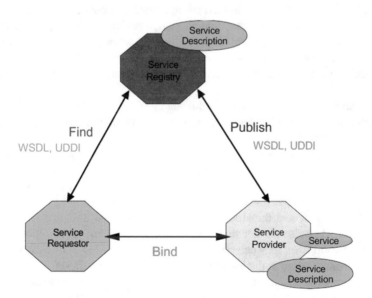

**Fig. 2.2** Web services roles, operations, and artifacts

## 2.4  SOA Adoption Case Studies and Surveys

Nowadays, SOA is the preferred architectural model to prepare the agility of organizations, to improve the adoption of applications and the interoperability of systems, and to permit the reuse of legacy systems [18]. The acceptance of service-oriented standards is not a simple process with the reason of creating service-oriented architecture. It contains a lot of projects in a long period of time. Once adopting SOA, dissimilar parts in a company are affected: organizational construction, people, workflow processes, and technologies [19]. It is worth mentioning that the required organizational redesign should be well organized like culture and individual behavior [4].

SOA adoption needs an important transformation in a business process philosophy and technology foundation. This attempt is determined by the promise of important benefits [2]. SOA adoption will prepare organizations by developing interoperability, legacy systems integration, reusability, organizational agility, composability, standardized data representation, and vendor-neutral communications infrastructure [20]. These attributes will lead to specific business advantages such as enhanced flexibility, improved speed to the marketplace, incremental deployments, and enhanced productivity were between the other probable benefits [21]. While the economic value provided by SOA drives from the acceptance of organizational decisions, this value is not assessable or even detectable [22, 23].

In a study, Chen [24] itemized the potential factors that influence on SOA adoption as the following: compatibility, visibility/observability, complexity, relative advantage, IT skills/expertise, IT architecture/infrastructure, trialability, company size and industry type, organizational culture, vendor support, financial cost, business partners' demand/readiness, tool support, standards maturity, and perceived benefits. In another research, Wu [25] identified the following factors as the influential factors in the way of SOA adoption in his study: compatibility, customizability, communicability, complexity, relative advantage, financial justification/cost, and visibility/divisibility.

Yoon and Carter [8] find out the following factors as potential factors in the way of migrating to SOA: (1) system integration, (2) IT and business alignment, (3) response to market changes and customer demands, (4) data flow, (5) customer service, (6) application development time and cost, (7) reuse existing, (8) applications, (9) operational cost, and (10) time to market. They stated that selecting and using SOA is influential on system integration, association between IT and business. It also influences the reaction to the alterations in market and customer demand, data flow, and the strategy through which organizations give service to their clients. They also pointed that SOA adoption could have an effect on the costs in an organization. So, they categorized the influence of using SOA from the standpoint of business value as the following two groups: cost and business agility.

According to Cherbakov et al. [26], how to advance and harmonize the IT system with business transformation is the most important question. Various approaches [4, 27] attempt to explain the complete process of SOA adoption. It is vital to declare that all recommended approaches cannot substitute with a company precise approach. In fact these proposed approaches can be helpful to make individual approaches. Sometimes adoption and implementation of an information technology are used spontaneously. For instance, Beimborn et al. [28] in a research about SOA business value drivers and deterrents express that organizations may only implement SOA, while they observe that SOA will increase their business value.

Joachim et al. [29] stated that the word adoption serves this purpose to make full application of an invention. The implication of "organizational adoption of an innovation," as the method of accepting a new idea, or technology, during a certain time is the result of reviewing the empirical research by Finch [30]. Finch further extends the concept, and states the application of technology includes mutual working of individual in a system that match in greater organizational process. Hence, technology adoption must be done within a given organization area.

Often organizations get in SOA projects with no proper up-front analysis of all the conceptions of their purposes [18]. IBM presented some of their knowledge about a lot of firms that adopted SOA and recommended different domain of SOA adoption challenges: program management, firm, technology, and governance. The most sophisticated of these areas are organization and governance, since they need the whole organization to change the styles, ways of communication, tool of collaboration, and different ways of reportage relationships [31].

While all different organizations can use SOA, it is considered that small- and medium-sized companies use the advantages of SOA more than other sizes, as it offers them a chance to present fee-based services [32]. In 2009, Forrester SOA Survey declared that adopting SOA is less in smaller firms which have less than 1000 employees [33]. SOA adoption in the manufacturing seems to be slower than preferred [34, 35]. Some industry informed that SOA failed to provide its promised advantages and is too costly as well [36]. Moreover, the elements in common IT project administration tend to engage all people who take financial benefits, start it with good plans to strengthen the basis, select and use increasing method, be backed by top managers and participate in personnel training course provided for SOA [37].

In spite of various models of the SOA adoption of individual organizations in the trade press [38, 39], it is too complicated to develop a precise representative description of SOA adoption [40] or a qualitative method [41]. A well-known exception is the study by Kumar et al. [42]; this research pays attention to inter-organizational profits. Ciganek et al. [37] tried to find out crucial factors that are challengeable for SOA adoption and web services by selecting eight organizations in different industries. They found out that many factors are standard for all organizations, whereas some factors are varying between industries. Hence, the steps and process of adoption can be different between various industries.

Kanchanavipu [43] believed that a SOA cannot be successful unless a SOA dominant power of a powerful authority is implemented in its real position in the organization. Kanchanavipu also stated that the powerful authority of SOA forms it and it is the person in charge who should decide about everything that determines anticipations and funding power as well as confirming behavior and achievements. Power of a powerful authority is essential in SOA, because the notion must be delivered to every one of the individuals who have to execute it or the act of creating a SOA has an impact on them. Trade clients, developers, architects, executives, and so on can be considered as this type of individuals. People may not find out the value of SOA by hearing some information about its nature during a long time. SOA should be controlled and monitored powerfully so that it can attract faithful clients with positive insights to help the organization achieve its goals.

The CA Wily [44] SOA adoption study results show that diverse countries get a different level of SOA adoption. A major number of organizations in Australia (32.9%) and the United States of America (40.6%) had applied a business unit service-oriented architecture application in IT control, while most of the organizations in the United Kingdom (40.6%) had applied a service-oriented architecture application as a part of an organization-wide initiative. The main number of the organizations in Germany (30.6%) and France (45.2%) had their SOA applications in the pilot level.

## 2.5    SOA Adoption Challenges and Problems

SOA adoption is not the same as establishing an application, which can be developed in a short period of time. Indeed, it is like a long journey for an organization to achieve significant benefits of implementing SOA. A few vendors believe that SOA can be rapidly and simply implemented with their products. However, SOA adoption is an evolutionary process, more than revolutionary one, in the general understanding in the industry [45].

On the other hand, SOA is not complete and without problems. Some declaration like "SOA is dead" [46] and "SOA is a failure" Kenney [47] could be found through the Internet. One of the main reasons for these declarations is that SOA is concentrated on developing design techniques to guide developers how to build services, but does not involve the run-time features of the service such as the way of managing and maintaining services. The link with business objectives cannot be made; aims cannot be specified and not be found whether goals are gained without proper management. Although some standards have emerged to provide management requirements [48], the standard SOA is not adequately equipped to describe them in a brief and consistent manner [49].

The perplexity of SOA and a lot of overlapping and competing specifications were specified as problematic areas for firms that adopted web services and SOA [35]. When the discussion comes to systems integration through SOA, the complexity can affect the number of physical resources required to address integration and thus it may affect the whole project implementation [50].

Adopting SOA is not simple and a lot of problems happen such as undeveloped standards and inadequate information [51]. Lewis [18] believed misunderstanding about SOA are: "(a) *SOA provides a ready-made architecture for a system that can be bought and implemented off the shelf* and (b) *Legacy integration is easily achieved*"

- Service-oriented architecture is including just technology and standards.
- If standards are applied, then interoperability will be guaranteed.

- Examining of service-oriented architecture applications is the same as the one that used for standard applications.

There are some arguments among firms about dissatisfaction for adopting service-oriented architecture, for example, the dearth of planning and clear business case, the dearth of information about available services, the dearth of standards, and the dearth of governance [52]. Some other difficulties of adopting SOA are the incomprehension of the differences between distributed architecture and SOA, establishing SOA through an old method, incomprehension the requirements of SOA implementation, and web service protection. Implementing SOA without using a development plan or a clear strategy, without adopting different standards and platforms, without adjusting SOA standards in a firm and without employing XML as an important foundation are the main reasons of failures in SOA architecture projects [2].

Each organization should know that adopting SOA cannot be the correct reply to grant their requirements. They should recognize that deploying SOA will not solve all of their problems. Some SOA myths can guide to a delusion about the entity of SOA and how it may help firms [53]:

SOA accommodates an absolute architecture. SOA is an architectural prototype or a new method of improving software but not a system architecture by its own.

Legacy systems can be simply accommodated inside SOA. One of attractive undertaking for a firm to accept SOA is to give complete approval in reconstituting its previous legacy system facilities and finding an important Return on Investment (ROI). The process of shifting legacy systems is not simple and routine all the time, because this shift can involve an enormous adaptation try to represent the legacy systems as services.

Employing web services do not certainly mean SOA implementation. Most of the SOA offered meanings refer to web services technology as a method to implement SOA. Indeed, SOA may not always be implemented by applying web services but by applying other technologies, such as DCE (Distributed Computing Environment) or CORBA.

By application of WSDL and XML interoperability between services advanced by diverse organizations is assured. Web services employ XML to arrangement information, assisting syntactic interoperability. Though, web services provide determined restrictions such as lack of stating semantic data. This means that no appropriate information about the service is supplied. Interoperability demands semantic along with syntactic agreements to prosper.

The development in SOA needs a cultural modify that affects all business sections inside a firm.

According to the report of Frost and Sullivan on Aug 2009, next to the different chances of marketplace in India, the marketplace of SOA is at early stage and it stayed untapped by sellers. Dhruv Singhal, Senior Director, Fusion Middleware Sales Consulting, Oracle India, pointed out:

*Today, the primary driver for SOA adoption is the business demand that forces enterprise data centers to deliver more with minimal resources. SOA adopts open standards to reduce integration costs, provide composite applications, and reduce custom coding through configuration and enable self-sufficiency for the end user. It is expected to be a critical business enabler rather than a mere IT tool.* (As cited in [54])

In John Crupi's study (as cited in [55]), the CTO of enterprise web services works at Santa Clara, claimed that a top-down method is needed to help SOA as an architectural style to flourish. Crupi believed that the BU has to possess the business drivers, processes, and use-cases. After that it is IT's duty to supply the BU demands and possess the definitions of service. Crupi disagree with using a "bottom-up" approach to SOA development, in which existing systems are simply wrapped using web services to create a service layer.

Like other application plans or architecture, quality may be different. It is true about SOA too. Although it is important to recognize the degree of changes that are necessary as well as the attitude to select and use SOA completely, there is always this possibility that some create wrong service-centered architectures unconsciously and unintentionally. According to Erl [2], the following are a number of frequent and usual mistakes.

Perhaps it is not clear that the sections belonging to organizations are the only most serious barrier that does not let them proceed to selecting SOA. But the experience that IT experts and technologists possess can cause a big number of SOA notions to rise. However, for many individuals in IT departments, SOA is just a technological notion. As a result, for experts in IT, SOA is just an identical medium for web services. Their mistake is obvious. The criticism in their thought causes misleading in technology [56].

In order to be able to select and use SOA, IT was exposed to some changes. The service-oriented change towards activeness and loose coupling needs an alteration in the old usual type of improvement (design-build-test-deploy-manage) to repeated methods to constant Service modeling. The movement to loose coupling needs a various method for estimating that causes the reliance on a single-vendor policy and application to decrease. For changing to get far from point-to-point mingling to compositional, process driven functions using services from a wide arrangement of benefits across the organizations, instead of system-specific silos, improvement and control methods based on service domains are needed [57].

As a result, it can be said that the main obstacles for selecting and using SOA are not caused by trade administration. In fact they originate from IT organizations that a very big number of individuals in IT organizations consider SOA just as a notion and a series of technologies and infrastructure to represent, protect, implement, and control services [58].

## 2.6    Factors Affecting the Adoption of SOA

In this section, a vast number of earlier studies were reviewed. About 18 papers resulted from reviewed papers which are focused on the effective factors in SOA adoption. A summary of all potential factors in SOA adoption is available in Appendix B as a table.

In a research, Lawler et al. [59], for contributing to effective management of SOA, analyzed business, procedural, and technical factors. The higher important factors resulted from a research project survey and professional case studies of three technology organizations. The higher important factors in business were allocated from (a) efficiency and flexibility benefits, (b) agility, (c) financial benefits, (d) customer demand, (e) market and regulatory differentials, (f) competitive, and (g) executive technology leadership. The most important factors in procedural analysis were distributed from (a) education and training, (b) knowledge exchange, (c) naming conventions, and (d) procurement of technology. The factors of technical area were allocated from (a) external SOA domain on projects, (b) external process domain on projects, (c) XML standard (d) business process management software, and (e) web services best practices.

As it is mentioned earlier, Luthria and Rabhi [60] explained the different aspect of SOA adoption from the technical and business vision. In this study, the factors influencing the organizational adoption of SOA mention as follows: (1) perceived value to the organization, (2) organizational strategy, (3) organizational structure, (4) organizational culture, (5) potential implementation challenges, and (6) governance.

Ciganek et al. [37] tried to find out crucial factors that are challengeable in the adoption of SOA using web services by investigating eight cases in four various industries based on TOE framework. The research team discovered many factors are standard for all organizations, whereas some factors are varying between industries. Hence, the steps and process of adoption can be different between various industries. Factors extracted from this study are: (1) business partner demand, (2) industry fragmentation and inertia, (3) management awareness and support, (4) availability of expertise, (5) return on investment, (6) industry leadership, (7) performance of services-based applications, (8) vendor support, and (9) security.

Galinium and Shahbaz [61] used the theoretical propositions strategy in their master thesis. The factors which are mentioned in this study as influential factors leading the migration of legacy systems into service-oriented architecture are extracted from literature reviewed and empirical data findings from five case studies including furniture, bank, engineering and airline companies in Europe. The authors classified the success factors into three categories containing technical, business, and both technical and business viewpoints. The success factors are: (1) business process, (2) budgeting and resources, (3) potential of the legacy applications (size and complexity, reusability factors, level of documentation), (4) architecture of the legacy systems, (5) close monitoring, (6) strategy of migration, and (7) governance. The following factors are specified as other factors influencing on SOA adoption in these studies: (8) information architecture, (9) dependence on commercial product, (10) testing, and (11) technical skills.

Antikainen and Pekkola [62] discovered factors influencing successful SOA implementation in an exploratory study by interviewing IT and business people from nine organizations which are pioneer in implementing SOA in Finland. This study identified four themes, containing 11 various factors that are relevant to business and IT alignment of SOA development. The themes and factors are: (1)

Organization culture and human resources (a) organizational maturity, (b) competences, and (c) SOA team; (2) Processes and methodologies (a) business driven development, (b) governance, and (c) business stakeholder participation; (3) Communication and artifact (a) common language, (b) strategy, and (c) enterprise architecture framework; and (4) Technology (a) rapid development tools and (b) process automation.

Chang and Lue [63] in an exploratory study itemized the risk factors affecting on the adoption of service-oriented systems in the order of strength of effect as follows: (1) insufficient technology planning, (2) lack of expertise, (3) ineffective project governance, and (4) organizational misalignment. Moreover, the result of this study demonstrated that (5) technology newness and (6) resource insufficiency are not critical risk factors to the adoption of service-oriented systems, but they both are valued high as a risk factor.

Vegter [10] examined critical success factors for SOA implementation in a study. The researcher selected the following three critical success factors according to previous studies: (1) complexity of SOA, (2) reusability of services, and (3) governing the adoption of SOA process. This study illustrated that service reusability is not as critical as it is indicated in some literatures. Moreover, the author clarified that SOA adoption will increase (4) interoperability, (5) federation, (6) vendor diversification options, (7) business and technology alignment, (8) ROI, (9) organizational agility, and reduce (10) IT burden.

Lee et al. [50] took the form of an exploratory study based on a review of 34 SOA literatures and 22 interviews, identifying 20 critical success factors for the successful adoption of SOA. This study classified CFS in six categories: (1) awareness, (2) strategy, (3) organizational management, (4) technology infrastructure, (5) governance, and (6) project management. Critical factors from the viewpoint of this research are: (1) deepening of enterprise-wide perception of SOA, (2) strengthening perceptions of SOA by sharing success story, (3) building strong support for enterprise-wide core human resources, (4) clear goal-setting, (5) step by step evolution planning with consideration of current capacity, (6) framing an organizational model for SOA management, (7) fostering a partnership culture between business and IT, (8) developing training planning, (9) generating standard definitions of SOA technology, (10) defining scope of technology application/security foundation, (11) standardization of business process, (12) putting in place of enterprise-wide architecture management system, (13) definition of SOA-based development methodology, (14) project team organization for SOA, (15) strengthening business service oriented design process, (16) strengthening communication within a project, (17) managing SOA policy processes, (18) establishing a service development/operation management process, (19) assessing performance of service processes, and (20) building an industry-wide foundation for SOA. Clear goal-setting based on business value is the most important critical success factor in implementing SOA as both literature and interviews ranked in this study.

Joachim et al. [64] developed a multi-dimensional SOA adoption construct based on TOE framework, which allows the authors to make the degree of SOA adoption measureable and comparable among different organizations. By evaluating the research model, the following results disclosed. In the technological view, (1) compatibility and (2) relative advantage are significant factors in SOA adoption, while the (3) costs in adopting SOA prevent the adoption. (4) SOA-related expertise of IT employees, (5) organizational size, and (6) support from top management are all important determinants in the organizational context, while (7) experience is the single most significant influencing factor in the whole model. The results show that in fact (8) competition is not a significant factor on the degree of SOA adoption in the environment context. Anyway, (9) management fad is another factor that effect on the degree of SOA adoption. It shows that an organization is also affected by the adoption results of competitors.

Aier et al. [65] focus on service orientation as a design paradigm for information systems engineering. This study considers information systems as the integrity of persons, business processes, and information technology that process data and information in an organization. To clearly identify CFS in this research, the body of literature from both the ERP and the EAI viewpoint were analyzed. The critical success factors that are mentioned in the study are: (1) integration strategy, (2) governance, (3) momentum resources and strategic importance, (4) culture and communication, (5) integration architecture and design, (6) characteristics of integration projects, and (7) transparency of design artifacts.

Caimei Hu [66] in a study introduces the web service technology standard system based on framework of TOE and analyzes the main factors affecting the adoption and diffusion of web service technology standards. Based on this study factors that influence on SOA adoption from the point of technology are: (1) advantages of web service technology standards, (2) the complexity of web service technology standards, (3) knowledge barriers in the adoption of web service technology standards, and (4) standards immaturity. From the viewpoint of organization, factors which influence on SOA adoption are: (1) technology capability of organization, (2) organization philosophy, and (3) organization scale. Environmental factors based on TOE framework that mentioned as influential factors in this study are: (1) industry concentration, (2) stakeholder, and (3) industry technical inertia.

Findikoglu [67] in a research proposed a conceptual model to show the success of adopting web services as a technological innovation. The researcher declared some important factors that effect on SOA migration in his study as follows: (1) security, (2) reliability, (3) agility, (4) efficiency and flexibility, (5) deployability, (6) organizations' size and scope, (7) centralization, (8) formalization, (9) interconnectedness, (10) complexity, (11) IT and business alignment, (12) governance, (13) ROI, (14) technological knowledge, (15) quality and availability of human resources, (16) competitive pressure, (17) regulatory influence, (18) dependent partner readiness, and (19) trust in web services.

Seth et al. [54] in a study reviewed articles and research work related to SOA from 2001 to 2011 and identified the factors that are relevant to SOA implementation. Based on this study factors influencing on SOA implementation are: (1) governance issues, (2) migration factors, (3) legacy systems integration, (4) change management, (5) resource competences, (6) security risk, (7) risk management, (8) challenges in scope understanding, (9) integration business and IT, (10) return on investment, (11) BPM and business agility, (12) user involvement and organizational commitment, and (13) training and teaching methodology.

MacLennan and Van Belle [68] emphasized factors significantly influencing SOA adoption in South Africa. The results of this study emphasize that (1) complexity and (2) cost are only important for SOA project success, whereas the following factors are significant factors for SOA adoption and SOA project success: (3) multiple standards and platforms, (4) compatibility with the EA, (5) top management support, (6) good governance and (7) strategy, (8) adequate human and financial resources, (9) vendor support for integration and development tools.

In a research, Basias et al. [69] developed an initial conceptual framework in order to categorize and examine procedural, business, technical, and human influential factors of SOA adoption in e-banking sector. Researcher identified 125 factors that might influence on SOA adoption in their study. After a deeper analysis they minified these factors to 16 significant factors that might influence SOA adoption in e-banking industry for study in a real e-banking situation. The 16 possible influential factors are: (1) strategy (long-term business plan), (2) goal (based on business value), (3) financial benefits (of SOA adoption), (4) return on investment (related to SOA adoption), (5) IT agility–business alignment (business, actors, processes and technology alignment through successful SOA adoption), (6) costs (for hardware, software, people), (7) communication (good communication between different departments), (8) risk (risk management related to SOA adoption), (9) culture (cultivate SOA friendly environment), (10) management (an overall plan related to regulations, security and strategic business alignment), (11) resistance to change, (12) security (hardware, software, confidentiality, privacy, data protection), (13) IT infrastructure (hardware, software, people), (14) fatigue (related to time and workload), (15) stress (related to time, workload and new requirements), and (16) staff (experience and training).

Emadi and Hanza [70] identified factors that lead to successful outcomes in SOA projects in their research. A meta-study was implemented on a number of related publications. According to this research, critical success factors in implementation of SOA are: (1) organizational culture and human resources, (2) process and methodologies, (3) technology, (4) SOA registry, (5) SOA governance, (6) top management support, (7) trust between business units, (8) IT infrastructure, (9) business/IT communication, (10) business processes and (11) management.

In another study in 2013, Koumaditis et al. identified a various number of CSFs influencing SOA implementation by critically reviewing the literature and identified individual factors that may form CSFs for SOA implementation in healthcare sector. The 18 CSFs that emphasized in this study are: (1) alignment, (2) clear goals form, (3) complexity, (4) cost, (5) culture, (6) enforce decision, (7) experience, (8) governance, (9) long-term planning, (10) measurement, (11) maturity identification, (12) project identification, (13) resources, (14) roadmap, (15) roles, (16) standards, (17) team, and (18) testing.

In an article recently published, managerial decision problems are researched in an SOA model. Choi et al. [71] concentrated on the SOA execution decision and the influence of SOA on IS agility and cost in this paper. Generally, they investigated the following factors as effective factors in this area: (a) external business environment, (b) business agility, (c) strategy for implementing, (d) encompassing marketplace competition, (e) IT infrastructure, (f) flexible IS architecture, (g) customer requirements, (h) complexity, and (i) cost.

Table 2.3 shows the influential factors of SOA and web service adoption which are concluded from literature review of this study. This table includes limitation and future work of each paper reviewed for this study. Table 2.4 represented the significant factors in SOA adoption that have more than four frequencies in previous studies.

## 2.7  SOA Adoption Frameworks and Models

Diffusion of Innovation Rogers [72] and Technology–Organization–Environment framework [73] are two theories that are frequently used to investigate the acceptance of IT adoption in organizations [74].

### 2.7.1  TOE Framework

The procedure of creativity and novelty in organization is more complicated. Usually, some people who can be the advocates of the latest opinion or disagree on it, engage in this procedure. And every one of them can influence the innovation-decision.

In The Processes of Technological Innovation, the structure of technology–organization–environment (TOE) framework is explained by Tornatzky and Fleischer [73]. They explained about all of the procedure of innovation—expanded from the improvement of creativity by engineers to the selection and execution of these creativities by users in the environment of a company. This structure reveals a section of the procedure. That is, it shows the way through which the setting in a company affects choosing and using as well as execution of a creativity and novelty. TOE structure is a hypothesis justifying three dissimilar factors that a scope in a company is able to affect the choice about selection in an organization. Such elements exist in the setting or scopes related to technology and in the setting or scope related to the environment. All of these elements have an effect on the creativity in terms of technology.

The structure of TOE was first shown in its primary shape. Then it was changed in some research about IT selection and use, giving an effective structure based on analysis. So the researchers who work on the selection and use and also integration of dissimilar kinds of IT innovation can use it efficiently. Some particular elements in three settings might be dissimilar in several research, yet the foundation of this structure is valid and reasonably hypothetical, with reliably and experientially comforted, and is mostly possible to be utilized to IS innovation [75].

**Table 2.3** Influential factors of SOA adoption

| Author/year | Context | Factors | Limitation | Future work |
|---|---|---|---|---|
| Chen (2005) | – | Tool support, standards maturity, IT architecture/infrastructure, vendor support, perceived benefits, firm size, IT skill and expert | – | The future researchers are recommended to do more in-depth research for recognizing the procedure of selection, use, and distribution of web-services for various kinds of applications and industries. Also, more hypothetical and experimental studies on the advantages of web-services and their influence of businesses can cause organizations to assess these services in terms of policy |
| Ciganek, Haines et al. (2005) | Financial industry | Visibility/observability, complexity, security, tool support, IT skills/ expertise, software development, financial justification/cost, business partners demand/readiness, industry inertia/fragmentation, standards maturity, IT architecture/ infrastructure | The results in this research are merely obtained during the work currently done on four organizations. In future, the elements different from these elements may be influential so that those which not observed effective be significant in future. Since this research is a case study, the results may not be true or applicable all around the world | In later studies, evaluation of the results is suggested to be considered. For example, this research can be repeated in different industries or a bigger scaled numerical research can be conducted according to the results obtained in this research. Also the researchers can analyze the qualitative data more precisely to show additional attitudes. Through a longitudinal method, one can consider and evaluate a company again later and find out the similarities and differences in the ideas about selecting and use decisions in present time and in the future |
| Lippert and Govindarajulu (2006) | Organization | Security concerns, reliability, deployability, organizations' size and scope, IT skills/expertise, perceived benefits, competitive pressure; regulatory influence; dependent partner readiness; trust in the web service provider | – | In order to assess this pattern, the scope of practical studies in different organizations is needed to be utilized. So that the other options interfering or moderating the variables on the association between the selection of web services and these variables are allowed |

| Ciganek, Haines et al. (2006) | Industry | Relative advantage, compatibility, complexity, visibility/observability, performance, security, standards maturity, IT skills/expertise, financial justification/cost, management awareness and support, business partners demand/readiness, Industry-inertia/fragmentation, vendor support | The results of the current research were merely obtained through currently working on eight organizations. So the elements that are proved to be influential now may not be effective in future or vice versa | In future studies, evaluation and validation of the results are essential. A method to do that is to improve a numerical study according to the results of the current research. In a longitudinal method, the organizations can be re-examined to find out different insights about selecting and using the system through actual adoption or non-adoption decision |
| Yoon and Carter (2007) | Different case studies | System integration, IT and business alignment, response to market changes and customer demands, data flow, customer service, application development time and cost, reuse existing, applications, operational cost, time to market | — | — |
| Lawler, Benedict et al. (2009) | Information system | Agility, efficiency and flexibility, financial cost, business client participation, competitive, market and regulatory differentials, customer demand, culture of innovation, organizational change management, executive sponsorship, executive business leadership, executive technology leadership | The viability of this research is limited due to the limited numbers of firms that used services in an SOE strategy. Although they used basic methods of SOA, it can be considered a very simple and insufficient experience in SOE strategy. So later studies can be conducted in a bigger in-depth scale. Yet, they may be limited by the privacy of each firm | The results show the significance of the methodology related to the administration of a genuine plan on SOA, while it is not as significant as technical and business elements that can be included in future research |

(continued)

**Table 2.3** (continued)

| Author/year | Context | Factors | Limitation | Future work |
|---|---|---|---|---|
| Lee, Shim et al. (2010) | Finance, heavy construction industries, and ICT | Awareness, strategy, organizational management, technology infrastructure, governance, and project management | This research directed the interviews just in Korea | This study suggested a systematic classification of CSFs, and investigating the feasibility of these CFS, and testing if they can be measured in contrast to an enterprise's activities or not |
| Joachim, Beimborn et al. (2010) | – | Compatibility, relative advantage, costs, expertise, organizational size, top management support, experience, competition, management fad | – | – |
| Findikoglu (2011) | – | Security, reliability, agility, efficiency and flexibility, deployability, organizations' size and scope; centralization, formalization, interconnectedness, complexity, IT and business alignment, governance, ROI, technological knowledge, quality and availability of human resources, competitive pressure, regulatory influence, dependent partner readiness, trust in web services | The structure or frame might be a guideline for selecting and using web services by a big number of organizations. Nevertheless, noticing that organizations are not similar in many aspects, the structure should be modified based on the setting or scope | – |

| Reference | Type | Factors | Limitations | Notes |
|---|---|---|---|---|
| Aier, Bucher et al. (2011) | | Integration strategy, governance, momentum resources and strategic importance, culture and communication, integration architecture and design, characteristics of integration projects, and transparency of design artifacts | This study has three limitations: first is the poor quality of the architecture and service design measurement in proposed model, next is the heterogeneity of the sample and last one is about the limited generality of the findings for IS design | This study showed that the size of company has not the important effect on SOA to be successful, so a deeper analysis is needed to discover this matter. There are only a few publications on service design guidelines and only as far as IT architecture is concerned |
| Seth, Singla et al. (2012) | Multiple case studies | Governance issues, migration factors, legacy systems integration, change management, resource competences, security risk, risk management, challenges in scope understanding, integration business and IT, ROI, BPM and business agility, user involvement and organizational commitment, training and teaching methodology | The limitation of this study is that the survey reviewed papers which published during 2000 to 2011 | – |
| MacLennan and Van Belle (2012) | Various industries (South Africa) | Relative advantage, complexity, compatibility, perceived benefits, top management support, good governance and strategy, adequate human and financial resources, vendor support for integration and development tools, reliability, security | The results of the research do not represent the whole population and organizations of the South Africa | The possibility for future work is a combination of quantitative and qualitative cross-sectional analysis. Gorgeous data could increase quantitative research results and permit revising of the SOA adoption research model |

(continued)

**Table 2.3** (continued)

| Author/year | Context | Factors | Limitation | Future work |
|---|---|---|---|---|
| Basias, Themistocleous et al. (2013) | E-banking | Strategy, goal, financial benefits, ROI, IT agility–business alignment, costs, communication risk, culture, management, resistance to change, security, IT infrastructure, fatigue, stress, and staff | This study surveys the factors that influence on SOA adoption in E-banking and it does not cover all in industries | In the plan for next studies, improvement and execution of the model and assessment with experimental results should be considered |
| Koumaditis, Themistocleous et al. (2013) | Healthcare | Alignment, clear goals form, complexity, cost, culture, enforce decision, experience, governance, long-term planning, measurement, maturity identification, project identification, resources, roadmap, roles, standards, team, and testing | The proposed conceptual model in this study was tested only through a single case study | To test the proposed model in a greater degree, further research should use other forms of strategies or multiple case studies |
| Choi, Nazareth et al. (2013) | Organization | External business environment, business agility, strategy for implementing, encompassing marketplace competition, IT infrastructure, flexible IS architecture, customer requirements, complexity, cost | – | – |
| Emadi and Hanza (2013) | – | Organizational culture and human resources, process and methodologies, technology, SOA registry, SOA governance, top management support, trust between business units, IT infrastructure, business/IT communication, business processes, and management | This study did not test in a large scale in a real world and stayed in a theoretical level | Investigating how a planned SOA project influence other organizational factors this study does not cover |

**Table 2.4** Potential factors influence on SOA adoption

| Year | | 2009 | | | | | | | 2010 | | 2011 | | | 2012 | | 2013 | | | | Factors frequency |
|---|---|---|---|---|---|---|---|---|---|---|---|---|---|---|---|---|---|---|---|---|
| Factor \ Author | | Lawler et al. | Luthria and Rabhi | Cigánek et al. | Galinium and Shahbaz | Antikainen and Pekkola | Change and Lue | Vegter | Lee et al. | Joachim et al. | Aier et al. | Caimei Hu | Findikoglu | Seth et al. | MacLennan and Van Belle | Basias et al. | Koumaditis et al. | Emadi and Hanza | Choi et al. | |
| 1 | Business and IT alignment | | | | | | √ | √ | √ | | | | √ | | | √ | √ | | | 6 |
| 2 | Business/IT agility | √ | | | | | | | | | | | | | | √ | | √ | √ | 4 |
| 3 | Communication | | | | | √ | | | √ | | | | | √ | | √ | | √ | | 5 |
| 4 | Competitive issues | √ | | | | | | | | √ | √ | | √ | | | | | | | 4 |
| 5 | Complexity of SOA technologies | | | | | | | √ | | | | √ | √ | | √ | | √ | | √ | 6 |
| 6 | Costs | | | | | | | | √ | √ | | | | | √ | √ | √ | | | 5 |
| 7 | Education and training | √ | √ | | | | | | | | √ | | | | | √ | | | | 4 |
| 8 | Governance | | √ | | √ | √ | √ | √ | √ | | | | √ | √ | √ | √ | √ | √ | | 12 |
| 9 | Human resource | | | | √ | | | | √ | | | | √ | | | | | √ | | 4 |
| 10 | IT infrastructure | | | | | | | | √ | | | | | | | √ | | √ | √ | 4 |
| 11 | Organizational culture | | √ | | | √ | | | | | √ | | | | | √ | √ | √ | | 6 |
| 12 | Organizational strategy | | √ | | √ | √ | | | √ | | | | | | √ | √ | √ | | √ | 8 |
| 13 | Resources sufficiency/competences | | | | | | √ | | | | | | | √ | √ | | √ | | | 4 |
| 14 | Return on Investment (ROI) | | | √ | | | | √ | √ | | | | √ | √ | √ | | | | | 5 |
| 15 | Security issues | | | √ | | | | | | | | | √ | √ | | √ | √ | | | 5 |
| 16 | Technical skills/expertise | | | √ | √ | | √ | | | | | | | | | | | √ | | 4 |
| 17 | Top management support | | | √ | | | | | √ | √ | | | | | | | | √ | | 4 |

Researcher will use TOE framework for this study because it is accomplished at the enterprise level and has been tested widely by IT researchers to investigate innovation adoption at organizational context, and the results are promising [76]. Besides, for empirical research TOE framework can continue to be used. According to previous studies, most of the researches about IT adoptions at enterprise level used this noticeable model. Moreover, TOE framework has been performed successfully in a vast number of studies. Table 2.5 shows some TOE-based studies.

Some researchers used only some part of the TOE framework to understanding varied IT implementations. For example Chong et al. [77] merged Diffusion of Innovation theory and TOE framework for assessing IT adoption in E-commerce industry. Researchers used just environmental and organizational context of TOE framework in their research. Zhu et al. [78] combined TOE framework and DOI theory in their study. They only focused on technical and organizational factors of TOE framework. Premkumar and Roberts [79] focused on the environmental and organizational factors based on TOE framework to evaluate significance of IT in modern association but didn't use technical factor of TOE. Tiago Oliveira and Maria F Martins [80] used TOE framework for IT adoption in E-business. They provided a framework including all TOE factors in technological and organizational context.

## 2.7.2  Previous SOA Adoption Models

Luthria and Rabhi [60] proposed a framework for studying the adoption and implementation of SOA in organizations practically. Figure 2.3 represents the conceptual framework of their study. According to proposed framework, authors recommended that adopting SOA in organizations could concentrate on factors affecting the determination of adopting SOA and factor influencing SOA implementation in an organization.

Nils Joachim [81] in a dissertation thesis provided a model to examine SOA adoption from business perspective based on TOE framework. Figure 2.4 shows the adopted model. The organizational level of adoption in this research model directed to make changes and developments in traditional models of adoption.

In another study, MacLennan [68] developed a conceptual model for SOA adoption based on DOI theory and TOE framework by reviewing a large number of SOA and IT diffusion literature. The provided model is shown in Fig. 2.5. This study clarified critical success factors for adoption of information system innovation [82] (Fig. 2.6).

## 2.8  Organizational Performance

Organizational performance contains the real output and consequences of a firm as surveyed against its contracted outputs or purposes. The performance of organization encircles three particular sections of organization outcomes. First, financial

**Table 2.5** TOE-based studies [74]

| Authors | IT adoption | Analyzed variable |
|---|---|---|
| Martins and Oliveira (2009) | Internet web site E-commerce | *Technological context*: technology readiness; technology integration; security applications |
| | | *Organizational context*: perceived benefits of electronic correspondence; IT training programs; access to the IT system of the firm; internet and e-mail norms |
| | | *Environmental context*: internet competitive pressure; web site competitive pressure; E-commerce competitive pressure |
| Liu (2008) | E-commerce | *Technological context*: support of technology; human capital; potential support of technology |
| | | *Organizational context*: management level for information; firm size |
| | | *Environmental context*: user satisfaction; E-commerce security |
| Pan and Jang (2008) | Enterprise resource planning | *Technological context*: IT infrastructure; technology readiness |
| | | *Organizational context*: size; perceived barriers |
| | | *Environmental context*: production and operations improvement; enhancement of products and services; competitive pressure; regulatory policy. |
| Teo et al. (2006) | Deployment of B2B | *Technological context*: unresolved technical issues; lack of IT expertise and infrastructure; lack of interoperability |
| | E-commerce | *Organizational context*: difficulties in organizational change; problems in project management; lack of top management support; lack of E-commerce strategy; difficulties in cost-benefit assessment |
| | | *Environmental context*: unresolved legal issues; fear and uncertainty |
| Zhu et al. (2006b) | E-business initiation | *Technological context*: technology readiness; technology integration |
| | E-business adoption | *Organizational context*: firm size; global scopes; trading globalization; managerial obstacles |
| | E-business routinization | *Environmental context*: competition intensity; regulatory environment |

(continued)

**Table 2.5** (continued)

| Authors | IT adoption | Analyzed variable |
|---|---|---|
| Oliveira and Martins (2010a) | E-business | *Technological context*: technology readiness; technology integration; security applications |
| | | *Organizational context*: perceived benefits of electronic correspondence; IT training programs; access to the IT system of the firm; internet and e-mail norms |
| | | *Environmental context*: web site competitive pressure |
| Lin and Lin (2008) | Internal integration of E-business | *Technological context*: IS infrastructure; IS expertise |
| | | *Organizational context*: organizational compatibility; expected benefits of E-business |
| | | *Environmental context*: competitive pressure; trading partner readiness |
| Pudjianto and Hangjung (2009) | E-government assimilation | *Technological context*: ICT expertise; ICT infrastructure |
| | | *Organizational context*: top management support; compatibility; extent of coordination |
| | | *Environmental context*: regulatory environment; competition environment |
| Oliveira and Martins (2011) | Technological innovation decision making | *Technological context*: availability; characteristics |
| | | *Organizational context*: formal and informal linking structure; communication process; size slack |
| | | *Environmental context*: industry characteristics and market structure; technology support infrastructure; government regulation |
| Lee et al. (2009) | Knowledge management systems | *Technological context*: organizational IT competence; KMS characteristics (compatibility, relative advantage, and complexity) |
| | | *Organizational context*: top management commitment; hierarchical organizational structure. *Environmental context*: with external vendors; among internal employees |
| MacLennan and Van Belle (2012) | SOA adoption | *Technological context*: use of standards and platforms, complexity, costs, technology implementation challenges, relative advantages |
| | | *Organizational context*: organization size, industry, perceived risks, IT skills/expertise, top management support, strategy and plan, governance, resources, perceived benefits |
| | | *Environmental context*: vendor support, industry pressure, IT media influence |
| Joachim (2012) | SOA adoption | *Technology context*: relative advantage, compatibility, costs, complexity |
| | | *Organizational context*: organization size, top management support, IT expertise |
| | | *Environment context*: management fad, management fashion, competition |
| Ciganek, Haines et al. (2009) | SOA adoption | Industry leadership, industry fragmentation and inertia, business partner demand, availability of expertise, justification and ROI, management awareness and support, performance of services-based applications, vendor support for SOA, and security |

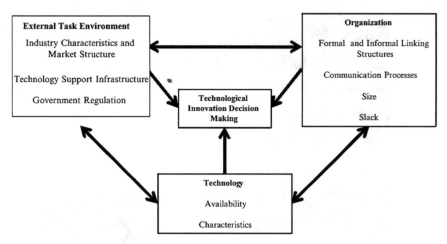

**Fig. 2.3**  Technology, organization, and environment framework

**Fig. 2.4**  Conceptual framework of the research agenda for the organizational adoption of SOA

accomplishment, such as return on assets, profits, returns on investment, and so forth. Second, product market performance such as market share, sales, and so forth. Third, shareholder return such as economic value added, total shareholder return, and so forth. The word organizational effectiveness is wider [83].

Few firms are able to assess their performance properly by averaging the accomplishment of their staffs. In most situations, the performance of a firm is established by the ability and efficacy of these upper-level organizational entities such as retail outlets, departments, teams, or plants. In the operations study, "decision-making

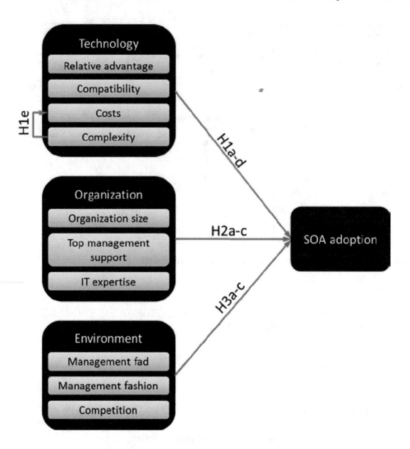

**Fig. 2.5**  Research model for investigating SOA adoption based on TOE framework

units" (DMUs) is the name of such productive entities. DMUs differs with each other when they use the same kind of sources and create the same form of outputs. Examples of DMUs throughout a firm are bank subsidiaries, shops in a chain of retail, or lines of assembly in a company [84].

## 2.9  Chapter Summary

As it is mentioned in the previous chapter, there is not enough knowledge about the effect of SOA adoption in organizations. According to this deficiency and the first objective of this study, researcher reviewed a vast number of previous studies related to SOA adoption and the effect of SOA adoption on organizations. Moreover, SOA definitions and benefits are described briefly. In the next step, web services as the best technologies for implementing SOA are explained and some standards related to web services are defined briefly like XML, SOAP, WSDL, and UDDI.

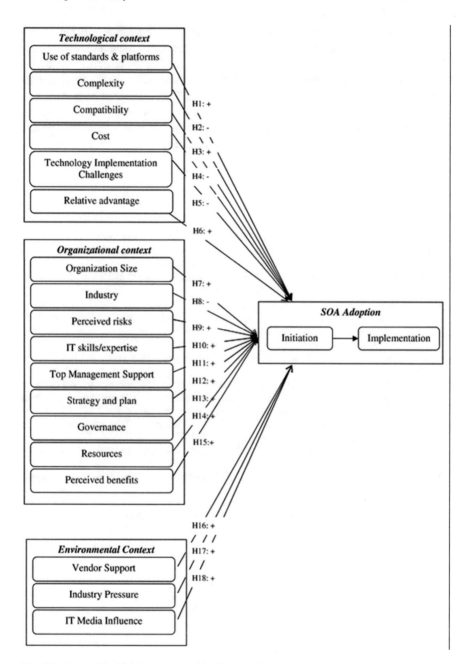

**Fig. 2.6** Research model for adopting SOA based on TOE framework

Various case studies about SOA adoption were reviewed and significant factors which effect on adopting SOA in organizations extracted from the reviewed papers. Based on the second objective of this study, for estimating the effect of key factors on SOA adoption and to evaluate the impact of SOA adoption on the performance of organizations, TOE framework is used. Based on TOE framework, this study will propose a framework to analyze the relations between SOA adoption and organizational performance which will be explained in Chap. 4. Therefore, three models due to the earlier studies reviewed at the end of this chapter.

# References

1. M. Matsumura, B. Brauel, J. Shah, *SOA Adoption for Dummies* (Wiley, 2009)
2. T. Erl, *Service-Oriented Architecture Concept, Technology and Design* (Pearson Inc., Upper Saddle River, 2005)
3. M.P. Papazoglou, W.-J. Van Den Heuvel, Service-oriented design and development methodology. Int. J. Web Eng. Technol. **2**, 412–442 (2006)
4. N. Bieberstein, S. Bose, L. Walker, A. Lynch, Impact of service-oriented architecture on enterprise systems, organizational structures, and individuals. IBM Syst. J. **44**, 691–708 (2005)
5. J. McGovern, O. Sims, A. Jain, *Enterprise Service Oriented Architectures: Concepts, Challenges, Recommendations* (Springer, Dordrecht, 2006)
6. A. Grigoriu, *SOA, BPM, EA, and Service Oriented Enterprise Architecture BPTrends*, www.bptrends.com, 2007
7. S. Jayashetty, P. Kumar, Adopting Service Oriented Architecture increases the flexibility of your enterprise, 2006. [Online]. Available: https://www.infosys.com/consulting/soa-services/whitepapers/Documents/adopting-soa-increases-flexibility.pdf. [Accessed: 29-Mar-2010].
8. T. Yoon, P. E. Carter, Investigating the antecedents and benefits of SOA implementation: A multi-case study approach, AMCIS Proc., Paper 195, 2007
9. E. Newcomer, G. Lomow, Understanding SOA with Web Services (Independent Technology Guides). Addison-Wesley Professional, 2004.
10. W. Vegter, Critical success factors for a SOA implementation, in *11th Twente Student Conference on IT*, Enschede, June 29th, 2009
11. C. Lawrence, Adapting legacy systems for SOA, Online, June, 2007
12. OASIS, *Reference Model for Service Oriented Architecture 1.0*, OASIS Standard, 2006
13. D. Booth, H. Haas, F. McCabe, E. Newcomer, M. Champion, C. Ferris, D. Orchard, *Web Services Architecture W3C Working Group Note 11 February 2004*, W3 Working Group, 2007
14. J. Bih, Service oriented architecture (SOA) a new paradigm to implement dynamic e-business solutions. Ubiquity **2006**, 4 (2006)
15. A. Skonnard, Understanding soap, MSDN Web Services Developer Center, 2003
16. IBM, Five best practices for deploying a successful service-oriented architecture, IBM Global Services, 2008. [Online]. Available: http://www-935.ibm.com/services/us/its/pdf/wp_five-best-practices-for-deploying-successfulsoa.pdf. [Accessed: 18-Apr-2013]
17. H. Kreger, *Web Services Conceptual Architecture (WSCA 1.0)* (IBM Software Group 5, 2001), pp. 6–7
18. G.A. Lewis, E. Morris, S. Simanta, L. Wrage, Common misconceptions about service-oriented architecture, in *2007 Sixth International IEEE Conference on Commercial-off-the-Shelf (COTS)-Based Software Systems (ICCBSS'07)*, IEEE, 2007, pp. 123–130
19. C.M. Pereira, P. Sousa, A method to define an enterprise architecture using the Zachman framework, in *Proceedings of the 2004 ACM Symposium on Applied Computing*, ACM, 2004, pp. 1366–1371

20. H. Demirkan, R.J. Kauffman, J.A. Vayghan, H.-G. Fill, D. Karagiannis, P.P. Maglio, Service-oriented technology and management: perspectives on research and practice for the coming decade. Electron. Commer. Res. Appl. **7**, 356–376 (2009)
21. L. Walker, IBM business transformation enabled by service-oriented architecture. IBM Syst. J. **46**, 651–667 (2007)
22. B. Mueller, G. Viering, C. Legner, G. Riempp, Understanding the economic potential of service-oriented architecture. J. Manag. Inf. Syst. **26**, 145–180 (2010)
23. R. O'Sullivan, T. Butler, P. O'Reilly, Realizing the business value of service-oriented architecture: the construction of a theoretical framework, in *Proceedings of the 6th European Conference on Information Management and Evaluation*, 2012, pp. 258–266
24. M. Chen, An analysis of the driving forces for Web services adoption. Inf. Syst. e-Bus. Manag. **3**, 265–279 (2005)
25. C. Wu, A readiness model for adopting web services. J. Enterp. Inf. Manag. **17**, 361–371 (2004)
26. L. Cherbakov, G. Galambos, R. Harishankar, S. Kalyana, G. Rackham, Impact of service orientation at the business level. IBM Syst. J. **44**, 653–668 (2005)
27. R. Schmelzer, J. Bloomberg, Zapthink's service-oriented architecture roadmap poster', Zapthink, 2005
28. D. Beimborn, N. Joachim, F. Schlosser, B. Streicher, The Role of IT/Business Alignment for Achieving SOA Business Value – Proposing a Research Model, AMCIS 2009 Proc., **335**, 1–8 (2009)
29. N. Joachim, D. Beimborn, P. Hoberg, F. Schlosser, Examining the Organizational Decision to Adopt Service-Oriented Architecture ( SOA ) - Development of a Research Model, Digit 2009 Proc. - Diffus. Interes. Gr. Inf. Technol., 2009
30. V. Finch, Company Directors: Who Cares about Skill and Care?, Mod. Law Rev. **55**, 179 (1992).
31. R. Varadan, K. Channabasavaiah, S. Simpson, K. Holley, A. Allam, Increasing business flexibility and SOA adoption through effective SOA governance. IBM Syst. J. **47**, 473–488 (2008)
32. D.K. Barry, *Web Services and Service-Oriented Architectures: The Savvy Manager's Guide* (Morgan Kaufmann, Burlington, 2003)
33. J. McKendrick, Forrester: Only one percent have negative experience with SOA, ZDNet, 2009. [Online]. Available: https://www.zdnet.com/article/forrester-only-one-percent-have-negative-experience-with-soa/. [Accessed: 24-Jun-2010]
34. K. Kontogiannis, G. Lewis, D. Smith, The landscape of service-oriented systems: a research perspective for maintenance and reengineering, in *Proceedings of the Workshop on Service-Oriented Architecture Maintenance*, Amsterdam, 2007
35. M.N. Haines, The impact of service-oriented application development on software development methodology, in *2007 40th Annual Hawaii International Conference on System Sciences (HICSS 2007)*, IEEE, 2007, pp. 172b–172b
36. M. Meehan, SOA adoption marked by broad failure and wild success, TechTarget, 2008. [Online]. Available: https://searchmicroservices.techtarget.com/news/1319609/SOA-adoption-marked-by-broad-failure-and-wild-success. [Accessed: 20-May-2012]
37. A.P. Ciganek, M.N. Haines, W.D. Haseman, Service-oriented architecture adoption: key factors and approaches. J. Inf. Technol. Manag. **203**, 42–54 (2009)
38. E. Schindler, Service-oriented architecture pays off for synovus financial, 2008
39. K. Flowers, Changing the game: how coca cola enterprises is leveraging SOA to transform their enterprise presentation at SOA summit 2009, Scottsdale, AZ, 2009
40. J. Löhe, C. Legner, SOA adoption in business networks: do service-oriented architectures really advance inter-organizational integration? Electron. Mark. **20**, 181–196 (2010)
41. D.W.-I.A. Becker, D.-W.-I.T. Widjaja, P. Buxmann, Value potentials and challenges of service-oriented architectures. Bus. Inf. Syst. Eng. **3**, 199–210 (2011)
42. S. Kumar, V. Dakshinamoorthy, M. Krishnan, Does SOA improve the supply chain? An empirical analysis of the impact of SOA adoption on electronic supply chain performance, in *2007*

*40th Annual Hawaii International Conference on System Sciences (HICSS 2007)*, IEEE, 2007, pp. 171b–171b

43. K. Kanchanavipu, An Integrated Model for SOA Governance rapport nr.: Report/IT University of Göteborg 2008: 002, 2008
44. C. Wily, CA Wily TechWeb study results, 2008
45. S. Mulik, S. Ajgaonkar, K. Sharma, Where do you want to go in your SOA adoption journey? IT Prof. **10**, 36–39 (2008)
46. A. T. Manes, SOA is dead; long live services, Burton Group, 2009. [Online]. Available: https://www.brighttalk.com/webcast/126/1722/soa-is-dead-long-live-services. [Accessed: 18-Mar-2010]
47. F. Kenney, Ahh Shucks, SOA Is A Failure, 2008. [Online]. Available: http://blogs.gartner.com/frank_kenney/2008/11/12/%0Aahh-shucks-soa-is-a-failure/. [Accessed: 14-Aug-2010]
48. OASIS, Web services distributed management: management of web services (WSDM-MOWS) 1.0, OASIS-Standard, March, 2005
49. M. Hiel, H. Weigand, W.-J. Van Den Heuvel, *Enterprise Interoperability III* (Springer, London, 2008), pp. 197–208
50. J.H. Lee, H.-J. Shim, K.K. Kim, Critical success factors in SOA implementation: an exploratory study. Inf. Syst. Manag. **27**, 123–145 (2010)
51. S. Tilley, J. Gerdes, T. Hamilton, S. Huang, H. Müller, D. Smith, K. Wong, On the business value and technical challenges of adopting web services. J. Softw. Maint. Evol. Res. Pract. **16**, 31–50 (2004)
52. M. Ren, K. K. Lyytinen, Building enterprise architecture agility and sustenance with SOA, Cais, **27 (4)**, 75–86 (2008)
53. L. Bastida, A. Berreteaga, I. Canadas, *Enterprise Interoperability III* (Springer, Berlin, 2008), pp. 221–232
54. A. Seth, A.R. Singla, H. Aggarwal, *Contemporary Computing* (Springer, Berlin, 2012), pp. 164–175
55. N. Bharti, Voices from the Web: SOA—top down or bottom up approach?, SearchSOA, 2005
56. L. Cherbakov, M. Ibrahim, J. Ang, SOA Antipatterns: The obstacles to the adoption and successful realization of service-oriented architecture, 2006. http://www.ibm.com/developerworks/webservices/library/ws-antipatterns/. Accessed: 01-Feb-2010
57. IBM Global services, Five best practices for deploying a successful service-oriented architecture (2008), http://viewer.media.bitpipe.com/1033409397_523/1212083066_235/wp_five-best-practices-for-deploying-successful-soa.pdf. Accessed 27 Nov 2012
58. K. Holley, J. Palistrant, S. Graham, Effective SOA governance IBM White Paper, IBM Corporation, March 2006
59. J.P. Lawler, V. Benedict, H. Howell-Barber, A. Joseph, Critical success factors in the planning of a service-oriented architecture (SOA) strategy for educators and managers. Inf. Syst. Educ. J. **7**, 1–30 (2009)
60. H. Luthria, F. Rabhi, Service oriented computing in practice: an agenda for research into the factors influencing the organizational adoption of service oriented architectures. J. Theor. Appl. Electron. Commer. Res. **4**, 39–56 (2009)
61. M. Galinium, N. Shahbaz, Factors affecting success in migration of legacy systems to service-oriented architecture (SOA), Master Thesis, Department of Informatics, Lund University, Sweden. Submitted, 2009
62. J. Antikainen, S. Pekkola, Factors influencing the alignment of SOA development with business objectives, in 17th European Conference on Information Systems (ECIS), 2009, pp. 2579–2590
63. H.-L. Chang, C.-P. Lue, *Designing E-Business Systems. Markets, Services, and Networks* (Springer, Berlin, 2009), pp. 83–95
64. N. Joachim, D. Beimborn, T. Weitzel, *Investigating Adoption Determinants of Service-Oriented Architectures (SOA) Sprouts: Working Papers on Information Systems*, 2010
65. S. Aier, T. Bucher, R. Winter, Critical success factors of service orientation in information systems engineering. Bus. Inf. Syst. Eng. **3**, 77–88 (2011)

66. C. Hu, *Computing and Intelligent Systems* (Springer, 2011), pp. 81–87
67. M. Findikoglu, *Web Services Adoption Process: A Roadmap to Manage Organizational Change Available at SSRN 1873723*, 2011
68. E. MacLennan, J.-P. Van Belle, Factors affecting the organizational adoption of service-oriented architecture (SOA). Inf. Syst. e-Bus. Manag. **12**, 1–30 (2012)
69. N. Basias, M. Themistocleous, V. Morabito, SOA adoption in e-banking. J. Enterp. Inf. Manag. **26**, 719–739 (2013)
70. S. Emadi, R.H. Hanza, Critical factors in the effective of service-oriented architecture. Adv. Comput. Sci. **2**, 26–30 (2013)
71. J. Choi, D.L. Nazareth, H.K. Jain, The impact of SOA implementation on IT-business alignment: a system dynamics approach. ACM Trans. Manag. Inf. Syst. **4**, 3 (2013)
72. E. M. Rogers, Diffusion of innovations, 5th ed. New York: The Free Press, 2003.
73. L.G. Tornatzky, M. Fleischer, A.K. Chakrabarti, *The Processes of Technological Innovation* (Lexington Books, Lexington, 1990)
74. T. Oliveira, M.F. Martins, Literature review of information technology adoption models at firm level. Electron. J. Inf. Syst. Eval. **14**, 110–121 (2011)
75. J. Baker, *Information Systems Theory* (Springer, 2012), pp. 231–245
76. C.H. Kok, P.L. Kee, A.L.H. Ping, S.N.A. Khalid, C.C. Yu, Determinants of internet adoption in Malaysian audit firms, in *2010 International Conference on E-business, Management and Economic IPEAR*, 2011, pp. 302–307
77. A.Y.-L. Chong, K.-B. Ooi, B. Lin, M. Raman, Factors affecting the adoption level of c-commerce: an empirical study. J. Comput. Inf. Syst. **50**, 13 (2009)
78. K. Zhu, S. Dong, S.X. Xu, K.L. Kraemer, Innovation diffusion in global contexts: determinants of post-adoption digital transformation of European companies. Eur. J. Inf. Syst. **15**, 601–616 (2006)
79. G. Premkumar, M. Roberts, Adoption of new information technologies in rural small businesses. Omega **27**, 467–484 (1999)
80. T. Oliveira, M.F. Martins, Understanding e-business adoption across industries in European countries. Ind. Manag. Data Syst. **110**, 1337–1354 (2010)
81. N. Joachim, *Service-Oriented Architecture (SOA): An Empirical Evaluation of Characteristics, Adoption Determinants, Governance Mechanisms, and Business Impact in the German Service Industry*, Department of Information Systems and Services University of Bamberg, 2012
82. E.M. Lennan, Factors affecting adoption of service-oriented architecture (SOA) at an enterprise level, Department of Information Systems, University of Cape Town, 2011
83. P.J. Richard, T.M. Devinney, G.S. Yip, G. Johnson, Measuring organizational performance: towards methodological best practice. J. Manag. **35**, 718–804 (2009)
84. G.A. Gelade, M. Ivery, The impact of human resource management and work climate on organizational performance. Pers. Psychol. **56**, 383–404 (2003)

# Chapter 3
# Quantitative Research Methodology for Service-Oriented Architecture (SOA) Adoption in Organizations

## 3.1 Introduction

Methodology is a series of procedures that is done to collect data to reach a definite purpose. There are two kinds of methodologies. The first one is qualitative data that is collected from interview, observation, books, and so forth. The next one is quantitative data that is collected from numbers and questionnaires. The main goal of this study is to identify the potential factors for the successful adoption of service-oriented architecture in Malaysian's organizations. In order to gain the purpose of this study, the researcher explains some parts, such as participants, research design, instruction, and so forth in this chapter. The participant's part introduces the subjects and the place where the data is collected. The next part introduces the design that is used for collecting data. The last part introduces how data is collected through instruments.

## 3.2 Research Design

An outline is necessary to arrange research activities step by step. The researcher in each step of study should understand where the study is and what the next step is. The study should organize the phases in the correct way. Quantitative analysis is used in this research. Figure 3.1 shows the research design.

© The Author(s), under exclusive license to Springer Nature Switzerland AG 2019     43
N. Niknejad et al., *The Impact of Service Oriented Architecture Adoption on Organizations*, SpringerBriefs in Electrical and Computer Engineering, https://doi.org/10.1007/978-3-030-12100-6_3

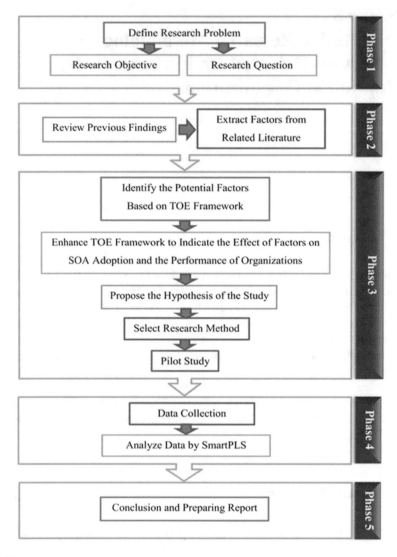

**Fig. 3.1** Research design

## 3.3  Research Method

There are different kinds of research methods that each researcher should select the proper method to achieve the main purpose of the study. All of them have their weaknesses and strengths. Table 3.1 shows the most important research methods. It is so important to choose an appropriate method for each task. For gathering the

**Table 3.1**  Different types of research methods [1]

| Research methods | Definition |
| --- | --- |
| Descriptive-qualitative (ethnography/case study) | Elaborate descriptions of particular case (Abdul Manan and Hyland) using interviews, observations, document review |
| Descriptive-quantitative | Description about numerical (frequency, average) |
| Correlational/regression analyses | Quantitative analyses between two or more variables about their strength of relationships (e.g., are qualifications of teacher correlated with achievement of students?) |
| Quasi-experimental | Without random sampling compare a group that receives a special treatment with the same group with the same characteristics that is not received |
| Experimental | Applying random sampling to determine subjects to an experimental group and a comparison or control group (e.g., one receives treatment and one does not) |
| Meta-analysis | Determining the average impact of the same intervention through studies by synthesizing the results of different studies |

data from people the interviews and questionnaires are used, whereas in document review several documents including papers, journals, business reports, etc. are used.

Data analysis techniques consist of statistical analysis and content analysis. These techniques will be selected according to the objectives and type of the research [2]. For SOA adoption studies, the quantitative method may provide a different comprehending. This technique will permit researchers to do their research with a vast view point in comparison with other data collection methods.

SOA adoption study has a various understanding in quantitative studies. In this regard, data condensing technique in quantitative studies permits researchers to see the vast view of different data collections elements; therefore, they can be suitable for researches that want to make a pattern of manner that relates to SOA adoption. Because it can measure the concepts and determine causality of variables, quantitative data analysis can be a useful tool to measure alignment by using strategy, virtues of culture of organization, and implementation challenges [3].

Since the focus of this study is on the companies which involved SOA, an adopted questionnaire is viewed as the appropriate option for this study for collecting data. So the researcher chooses the questionnaire to achieve the objectives of this study. Moreover, it seems a good method for gathering reliable and accurate data. So, a questionnaire has been selected as the best method for collecting data for two reasons:

The type of data that the researcher looks for is not available easily on papers. Thus, this research needs professional people who have adequate knowledge about SOA for answering the items of the questionnaire. Moreover, the position of these people in an organization is so important because it can definitely affect their decision.

The researcher needs to have several opinions of the people. These data in terms of the questionnaire is possible to have the best view of the situation. Using questionnaire is one of the quickest methods of conducting information from diverse groups of people.

## 3.4   Research Phases

It is very essential to have an academic review on service-oriented architecture and its influence on SOA adoption, based on TOE framework. The necessity of having a quantitative method to achieve the goal of this study is needed to lead the study to adopt a questionnaire to collect data for evaluating the impact of SOA adoption in organizations. The following parts describe the research phases.

### 3.4.1   Phase 1: Planning Phase

The first phase of this study is the planning phase. In this phase, the purpose and objectives of the study are identified and statement of the problem is recognized. To achieve this goal primarily the previous results of other academic studies were reviewed and a proper scope was chosen. This phase consists of the proposal and the context of Chap. 1 of this study. The following steps show this phase in detail:

To define research problem and the background of the study a large number of studies about SOA adoption in organizations were reviewed. Based on previous studies and the aim of this study, TOE framework is selected as a framework to identify the impact of SOA adoption on organizations.

To identify the objectives of the study and research questions, the author reviewed the academic studies about SOA adoption and the impact of SOA on organizations deeply. The goals of this study are determined based on problem statement and background of the study. To select the scope of the study the researcher chooses the organizations which adopted or implemented service-oriented architecture.

### 3.4.2   Phase 2: Review of the Literature

In this phase, a vast number of academic papers were reviewed which had been published in the high impact factor journals and conferences during 2009 and 2014. It is worth mentioning that online database search engines such as Google Scholar, IEEE Xplore, Springer, Science Direct, and Elsevier were used in this study.

In the first step, the keywords like SOA adoption and SOA implementation in organizations were used and about 27,700,000 papers were found in the research result. Then the papers restricted to those which investigated the factors affected SOA adoption and SOA implementation in organizations from 2009 till 2013. The keywords used for this step were: influential factors, success factors, critical success factors, and factors affected SOA adoption. In the next step of this phase, researcher reviewed the selected papers deeply and the potential factors which are influenced on SOA adoption extracted from previous studies.

### 3.4.3 Phase 3: Data Collection Method

Data collection is the process that researcher gather necessary information from different participants. The result of the study is based on the data collection. As it is mentioned before there are two main data collections. The first one is primary data that is based on the researcher's own experience including observations, question-naires, interviews, and so on while secondary data is based on other researchers' experiences and researches. At qualitative method in common researchers should make an appointment with the target group of the study. The most significant chal-lenge in this study is the limitation of contacting with companies. Due to this rea-son, the author could not use the qualitative method, therefore the quantitative method has been chosen for collecting data in this research.

**Target Population and Sampling**

The preliminary aim of this study was to do a research on Malaysian organizations which are adopting and implementing SOA. But contacting with such organizations was the huge challenge of this study. The researcher does not access to any mailing lists of Malaysian organizations. However the author made a connection with one of the leaders of SOASchool, SOA Education Inc., to find a list of SOA organizations in Malaysia. It seems too costly and time consuming to provide a list of SOA adopted organizations. So, the researcher tried to connect with those companies which are available through Internet and found some companies which are adopted to SOA. None of those companies accepts to participate in this study. Finally, according to the target groups of this research which are IT experts and SOA profes-sionals and since the researcher does not have enough knowledge about the target group who are able to participate in this research, it was decided to use non-probabilistic sampling and self-selecting technique (Table 3.2).

Non-probability sampling is a technique that does not provide the equal chance for all individuals in the population to be selected [2]. One type of non-probability technique is self-selection sampling. It is useful when a researcher allows organiza-tions and individual choosing to take part in a research on their own agreement [4]. In this method, an online questionnaire will be distributed among individuals and organizations. So, using non-probabilistic sampling in this study is not going to be the indicator for all organizations all over the world.

As it is mentioned before, the target population of the survey should be chosen in this section. For choosing the most appropriate respondents two things must be taken into account. First, this study is about the organizations which adopted SOA. Second, its aims are to identify factors which influence on SOA adoption in organizations and to deal with the barriers in adoption of SOA. It indicates that the respondent should have adequate knowledge on information technology espe-cially service-oriented architecture issues. As the result, only experts including IT

**Table 3.2** Types of internet-based surveys and associated sampling methods [4]

| Sampling method | Web | E-mail |
|---|---|---|
| **Probability-based** | | |
| Surveys using a list-based sampling frame | √ | √ |
| Surveys using non-list-based random sampling | √ | √ |
| Intercept (pop-up) surveys | √ | |
| Mixed mode surveys with Internet-based option | √ | √ |
| Pre-recruited panel survey | √ | √ |
| **Non-probability** | | |
| Entertainment polls | √ | |
| Unrestricted self-selected surveys | √ | |
| Surveys using "harvested" e-mail lists (and data) | √ | √ |
| Surveys using volunteer (opt-in) panels | √ | |

managers or IT staff of organizations can participate and help in filling out the questionnaires. The following classification was used for target of the study:

- CIO, CTO, Chief Technical Architect, CSO/CISO, VP of IS/IT
- IS Manager, Director, Planner
- IS/IT/Technical Architect
- Other IT Manager in IS Department
- IT Staff

As it is described before, the respondents of the questionnaire in this study are IT managers and IT staffs who have adequate knowledge about SOA. The researcher made connection with SOA professionals through LinkedIn by sending an e-mail to them separately. The context of sampling letters in order to make connection through LinkedIn is available in Appendix B.

LinkedIn is one of the biggest social networking websites which is suitable for people in professional occupation. LinkedIn reports in June of 2013, more than 259 million members use this social network in over 200 countries and regions and it is available in at least 20 languages [5]. The countries with the highest LinkedIn users in January 2013 were [6]:

- United States = 74 million members
- India = 20 million members
- United Kingdom = 11 million members
- Brazil = 11 million members
- Canada = 7 million members
- Australia = 5 million members
- UAE = 1.3 million members

Moreover, LinkedIn has an advanced search tool. Members may search based on:

- People
- Jobs
- Companies
- Groups
- Universities
- Inbox
- All

By selecting *"People"* in the search tool and using the keyword *"SOA profes-sionals"* a list of experts were resulted. Then the link of the online questionnaire was sent to each expert with a message text individually. The message consists of an explanation about the survey and the purpose of the study. The content of this mes-sage is available in Appendix B. In addition, the researcher used an excel file as a database to save the name of each profession that the questionnaire was sent to them and any feedbacks received from the experts saved in this file. The questionnaire was sent totally to 369 potential respondents. Every week a friendly reminder was sent to experts and encourages them to take part in research and fulfill the question-naire. An overall number of 117 questionnaires were collected. Three of 117 ques-tionnaires were filled via SOA researchers, so these three questionnaires were considered as unsuitable for data analysis. Finally, 104 responses were acquired as a result of data sampling.

Since the number of target population was not clear, so it is slightly difficult to estimating the response rate. According to Oates [7], the rate of the questionnaire response should be from 10 to 30%. Due to the data available to the author, the rate of the questionnaire response can be predicted at 28.2% (104/369).

- **Validity**

- The questions were primarily considered with the supervisor and co-supervisor, so some questions refined. Moreover, two experts confirm the validation of the questions. As many times it mentioned before, the respondents of this study are selected from two large groups in LinkedIn, *SOA Professional Worldwide* and *Arcitura IT Certified Professionals*. To select appropriate person for this study researcher peruse the profile of each expert individually to be sure that they have more than a year experience in SOA issues and most of them were certified as a SOA architect.

- **Reliability**

- The best common factor for measuring reliability of a questionnaire is Cronbach's alpha. According to [8] for exploratory studies the minimum level of acceptance for reliability is 0.6 [9, 10]. In this study, a pilot study was conducted for evaluat-ing the reliability of the questions. The response from analyzing the pilot test was mainly positive. Just some questions were rephrased after researcher received the feedback from the pilot testing.

### 3.4.4    Phase 4: Data Analysis

In this study, the researcher used both primary and secondary data. Most of the secondary data was collected in review of literature part. And the primary data is the questionnaire that will be explained in the next chapter deeply. Moreover, it is concluded that path analyzing is an appropriate way to test a framework and hypothesis by reviewing previous studies. SmartPLS is one of the common path modeling tools and it can help researchers to save their time and analyze data more quickly and easily. Besides, researchers can achieve accurate predictions and organize research's outcomes.

Moreover, this study established nine hypotheses which were proposed and evaluated with SmartPLS software. The hypotheses are represented in Table 3.3. These hypotheses covered the second research question of this study which supposed to clarify the relationships between key factors, SOA adoption, and the performance of organization.

Since this study proposed a framework so, SmartPLS is chosen to determine the relationship between dependent and independent variables. In order to identify the effect of factors on adoption of service-oriented architecture in organizations and the impact of SOA adoption on the performance of organizations, the Smart Partial Least Squares (SmartPLS) version 2.0.M3 for MS Windows 7 is used as a quantitative analysis.

### 3.4.5    Phase 5: Conclusion and Preparing Report

In this phase, all findings and results from analyzing data are explained in detail. Moreover, based on the objectives of the research about significant factors and the impact of SOA adoption on the performance of organizations, some recommendations for achieving the highest benefits of SOA are expressed.

**Table 3.3**  The proposed hypothesis of relationship between variables in this study

| | |
|---|---|
| H1 | The complexity of SOA technology negatively influence on the adoption of SOA |
| H2 | The security concerns negatively influence on SOA adoption |
| H3 | Costs negatively influence on SOA adoption |
| H4 | SOA governance positively affects adopting SOA in organizations |
| H5 | SOA strategy positively influence on SOA implementation |
| H6 | Culture and communication positively influence on adopting SOA |
| H7 | Business and IT alignment positively influence on SOA adoption |
| H8 | Return on investment positively affects SOA adoption |
| H9 | SOA adoption positively influence on the performance of organization |

## 3.5    Chapter Summary

This chapter described the methodologies that are used in this study. This research consists of five phases: (1) planning phase, (2) reviewing previous studies, (3) extracting factors and proposing theoretical framework, (4) collecting and analyzing data, and (5) preparing report and conclusion. Due to the limitation of the study, researcher used self-selected sampling for this research and questionnaire distributed among IT experts who have adequate knowledge and experience about SOA through LinkedIn. Besides, for analyzing data and evaluating the proposed framework SmartPLS is chosen.

## References

1. SERVE, *Types of Research Methods* (University of North Carolina at Greensboro, Greensboro, 2008)
2. J. J. Castillo, Non-probability sampling, Explorable.com, 2009. [Online]. Available: https://explorable.com/non-probability-sampling. [Accessed: 08-Aug-2011]
3. H. Luthria, F. Rabhi, Service oriented computing in practice: an agenda for research into the factors influencing the organizational adoption of service oriented architectures. J. Theor. Appl. Electron. Commer. Res. **4**, 39–56 (2009)
4. R. Fricker, Sampling methods for web and e-mail surveys, in *The SAGE Handbook of Online Research Methods*, ed. by N. Fielding, R. M. Lee, G. Blank, (Sage, London, 2008), pp. 195–216
5. J. Hempel, *LinkedIn: How It's Changing Business (And How to Make It Work for You) Fortune*, vol. 168 (2013), pp. 68–74
6. D. Nishar, *200 Million Members!, LinkedIn Blog*, January 2013
7. B. Oates, *Researching Information Systems and Computing* (Sage, London, 2006)
8. J. C. Nunnally, Psychometric theory. New York: McGraw Hill, 640 Pages. 1967
9. E.M. Lennan, Factors affecting adoption of service-oriented architecture (SOA) at an enterprise level, Department of Information Systems, University of Cape Town, 2011
10. E. Ngai, T. Cheng, S. Ho, Critical success factors of web-based supply-chain management systems: an exploratory study. Prod. Plan. Control **15**, 622–630 (2004)

# Chapter 4
# Developing of Service-Oriented Architecture (SOA) Adoption Framework and the Related Hypotheses

## 4.1 Introduction

The main effort of this chapter is to test the initial study. Concluded factors from literature review of the study can guide to categorize the fundamental domain of research and will be used as primary definitions for next step of the research. In this chapter, the relationship among significant factors, SOA adoption and the impact of them on the performance of organizations are discovered. Selected factors are extracted from reviewing the previous studies. Moreover, hypothesizes are investigated and a questionnaire is designed to validate the positive effects of factors based on TOE framework to evaluating the adoption of SOA on the performance of organizations. To aim the goal of this study a proposed TOE framework is developed in the next step. At the end of this chapter, the questionnaire is analyzed to prove the reliability and validity of the questions in pilot study to start the analyzing step for the next chapter.

## 4.2 Proposed Framework

In this section, a framework is recommended by eight constructs which influence on SOA adoption and performance of organizations. As it is mentioned before, these eight factors are extracted from previous studies. Table 4.1 shows selected factors based on literature reviewed in this study. It was identified that there could be a relationship between these factors and SOA adoption and the performance of organizations.

According to Chap. 2, the most influential and significant factors that concluded from reviewed papers are: (1) complexity, (2) security concerns, (3) costs, (4) culture

© The Author(s), under exclusive license to Springer Nature Switzerland AG 2019
N. Niknejad et al., *The Impact of Service Oriented Architecture Adoption on Organizations*, SpringerBriefs in Electrical and Computer Engineering, https://doi.org/10.1007/978-3-030-12100-6_4

**Table 4.1** Influential factors on SOA adoption based on TOE framework

| Year | | 2009 | | | | | | 2010 | | | 2011 | | | 2012 | 2013 | | | | | |
|---|---|---|---|---|---|---|---|---|---|---|---|---|---|---|---|---|---|---|---|---|
| TOE | Author / Factor | Lawler et al. | Luthria and Rabhi | Ciganek et al. | Galinium and Shahbaz | Antikainen and Pekkola | Change and Lue | Vegter | Lee et al. | Joachim et al. | Aier et al. | Caimei Hu | Findikoglu et al. | Seth et al. | MacLennan and Van Belle | Basias et al. | Koumaditis et al. | Emadi and Hanza | Choi et al. | Factors frequency |
| T | 1 Complexity | | | | | | | ✓ | | | | ✓ | ✓ | ✓ | ✓ | | | | ✓ | 6 |
| | 2 Security concerns | | | ✓ | | | | | ✓ | | | | ✓ | ✓ | | ✓ | | | | 5 |
| | 3 Costs | | | | | | | | | ✓ | | | | | ✓ | ✓ | ✓ | | ✓ | 5 |
| O | 4 Culture and communication | ✓ | ✓ | | | ✓ | | ✓ | ✓ | | ✓ | | | | | | | ✓ | | 7 |
| | 5 Governance | ✓ | ✓ | | | ✓ | ✓ | ✓ | ✓ | | ✓ | | ✓ | ✓ | ✓ | ✓ | ✓ | | | 12 |
| | 6 Strategy | ✓ | ✓ | | ✓ | ✓ | ✓ | | | | | | | | ✓ | ✓ | ✓ | | | 8 |
| | 7 Business and IT alignment | | | | | | ✓ | ✓ | ✓ | | | | ✓ | | | ✓ | ✓ | | | 6 |
| | 8 ROI | | | ✓ | | | | ✓ | | | | | ✓ | ✓ | | ✓ | | | | 5 |

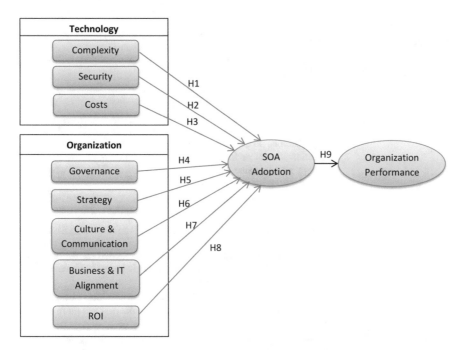

**Fig. 4.1** The enhanced framework for measuring the impact of SOA adoption on the performance of organizations

and communication, (5) governance, (6) strategy, (7) business and IT alignment, and (8) return on investment. Due to the reviewed papers for this study, these factors have more popularity since 2009. Table 4.1 shows resulted and categorized factors based on TOE framework.

Based on Table 4.1, the proposed framework for this study is shown in Fig. 4.1. As it is clear in this framework, all factors have relationships with SOA adoption and adopting SOA in an organization will affect the performance of the organization. In the following table (Table 4.2), the factors this study considered are briefly described.

Figure 4.1 represents the proposed framework of this study. As it is clear, this study has tried to evaluate relationships among key factors, SOA adoption and organization performance. Since, in the selected factors there are some performance factors like ROI, business and IT alignment, and culture and communication; hence the researcher has attempted to measure the effects of all factors together on the performance of the whole organization.

## 4.3 Hypothesis Development

According to Chap. 2, the literature review of this study, eight factors were identified. This section explained the relationship between these factors and SOA adoption and the effect of them on the performance of organization. In the next step, the

**Table 4.2** Factors description

| | Factors | Definition |
|---|---|---|
| 1 | Complexity of SOA technologies | Complexity is the degree to which a certain innovation is difficult to understand and use [1] |
| 2 | Security concerns | Security is fundamentally about protecting assets. Assets may be tangible items, such as operations or your customer database or they may be less tangible, such as your company's reputation |
| 3 | Costs | Cost is a vital factor related to any project and in any sector. Usually organizations do not proceed to the implementation of a software solution without performing a detail examination of all cost types and categories associated with their project. The same happens when organizations implement SOA [2] |
| 4 | Culture and communication | Organizational culture supports organizations to integrate their IT infrastructures. Culture and communication provide a SOA friendly environment [3] |
| 5 | Governance | SOA Governance is defined as a management model or a governance plan that provides compliance with internal/external regulations and checks services concerning security, strategic business alignment, and capability [3, 4] |
| 6 | Strategy (long-term planning) | SOA pays attention to reusability of services, so a long-term plan is needed that contains reusable services to make business more easily in future [4] |
| 7 | Business and IT alignment | Business and IT alignment is clarified as a degree that plans, purposes, and the mission enunciated in strategy of business are shared and advocated by using IT strategy. The capacity of determining a positive correlation between financial evaluation of performance and IT is alignment ([5], [6]) |
| 8 | Return on Investment (ROI) | To measure the ROI, the benefit of an investment is divided by the cost of the investment. The outcome is represented as a ratio or a percentage |

relations between the components of the framework are shown with hypotheses. This study consists of eight independents and two dependent constructs.

### 4.3.1   Complexity

According to Vegter [7], complexity is frequently cited as a reason for SOA projects to fail. So it can be concluded that it is one of the significant factors that influenced by adopting SOA in organizations. Moreover, based on Hu [8] declaration complexity is the complication that relates to comprehension and utilizing a technology that referred to the level of adoption negatively. Most of firms suppose that they may adopt the technology of web services when it's necessary even though the technology of web services standards are sophisticated a lot.

In the discussion of implementing SOA, the complexity influences many physical resources that need to be integrated and so it will affect the whole implementation of the project. Therefore, it is concluded that complexity negatively influences SOA adoption and the following hypothesis is proposed:

**H1:** The complexity of SOA technology negatively influence on the adoption of SOA.

### 4.3.2   Security Concerns

During SOA adoption, protection of information and business continuity risk have been the important topics that should be handled. Security of information is to guarantee that information security weaknesses and events are highlighted on time [9].

Security is an issue that covers the whole participating organization. It contains integrity, confidentiality, and the available exchangeable information via web services. Precisely, the important principle for wider adoption of web services and SOA as well is web services security standards' perceived maturity. This hypothesis would be concluded [10].

**H2:** The security concerns negatively influence on SOA adoption.

### 4.3.3   Costs

In an online report, Jeff [11] has declared that nobody can ignore the significant necessities for careful costing when an organization wants to migrate to SOA. Adopting SOA may take a long period of time to finish and may cost massive amounts of funds but SOA can be cost saving through making proper plan for budgeting. Without a good vision and anticipation on the essential budget for migration, companies would shift towards wasting huge amount of money on areas that are not the true subject to provide a long term for achieving whole SOA benefits [12]. From the above statements the following hypothesis is resulted.

**H3:** Costs negatively influence on SOA adoption.

### 4.3.4   Governance

According to Vegter [7], IT efforts are directed by existence of IT governance to confirm that the operation of IT comes through business objectives by the following way: alignment of IT with the business causing in understanding promised advantages, using occasion and maximizing the use of positive advantages, use of IT resources in a correct way and controlling the risks of IT related issues. So, SOA governance could be considered as a progress of IT governance presenting business contribution to providing the components of IT services.

Moreover, Mac Lennan [13] stated that one of the most important features of SOA adoption process is SOA governance. It is deliberated in organizations to be vital for understanding the profits of technology adoptions. Therefore, the following hypothesis is proposed for this study.

**H5:** SOA Governance positively affects adopting SOA in organizations.

### 4.3.5  Strategy

According to Erl [14], performing service-oriented architecture without a clear strategy and planning is one of the main reasons that caused SOA projects failing. Mac Lennan [13] demonstrated in a study that an effective SOA strategy in organizations will bring a superior potential for adopting SOA. The author illustrated aligning existing IT strategy and business with SOA strategy is essential for implementing SOA. So, it is believed that the following hypothesis would be accepted in this study.

**H6:** SOA Strategy positively influence on SOA implementation.

### 4.3.6  Culture and Communication

According to Aier et al. [3], the tradition of IS design prototype is broken with service orientated architecture essentially. So, it is vital for organizations to foster the culture of employees for willing to such change. Effective service orientated architecture will be achieved through the culture and communication supporting an essential change.

In another study, Emadi and Hanza stated that organizational culture assists to align business necessities in a SOA project. Accessibility of data and knowledge sharing and strategic plans between IT department and business department is another significant aspect of organizational culture factor. It is resulted that the culture of willingness for change is vital in migrating from tradition systems to SOA surroundings. Therefore, the below hypothesis is proposed:

**H4:** Culture and communication positively influence on adopting SOA.

### 4.3.7  Business and IT Alignment

According to Antikainen and Pekkola [4], SOA as IT strategy needs extraordinarily great level of business and IT alignment to attain benefits. Appropriate alignment is not achieved even a successful SOA project fulfills the technical objectives while the business part does not achieve the proper purposes.

Findikoglu [15] in a study disclosed that technological challenges are measured less important than organizational ones, forasmuch as both area are affected as a result of implementing web services. Hence, aligning business necessities and IT ability should be the main effort during the whole process of adopting SOA. Thus, the following hypothesis would be resulted.

**H7:** Business and IT alignment positively influence on SOA adoption.

### 4.3.8   Return on Investment

It is proper to embed the investment of technology and IT development in the same financing mechanism like other investments such as facilities while technology is considered as an investment, too. Most of the time management views all the investments in the same way. In a business developing precise value to express the ROI is crucial to success the technology development and IT function. The conclusive level of measurement is ROI. In this level, the cost of solution and financial benefits of solution are compared with each other [16].

An IT executive declared in a CIO Magazine article, "ROI has more credibility when it's stated in raw benefits, which are sometimes non-quantifiable, rather than translated into dollars. That translation is often fuzzy and tends to lose some audiences." Obviously, measuring technology ROI is accompanied by difficulty, but it is not impossible. In some way, if it is done correctly it could return a valued vision [17]. So the below hypothesis is proposed for this research.

**H8:** Return on Investment positively affects SOA adoption.

### 4.3.9   Correlation of SOA Adoption and Organization Performance

According to Gelade and Ivery [18], only few firms are able to assess their performance efficiently by averaging to the employees' performance. Productiveness and effectiveness of higher level entities of organization such as retail outlets, departments, plants, or teams determine the organization's performance. The purpose of this research is to evaluate the SOA adoption affection of the performance of organization; therefore, this hypothesis is designed. The objective of this study is to evaluate the effect of SOA adoption on the performance of organization. Hence, the following hypothesis is proposed.

**H9:** SOA adoption positively influence on the performance of organization.

## 4.4  The Study Instrument

As it is mentioned in the previous chapter, a structured questionnaire is selected as the most suitable research method. Therefore, an adopted questionnaire according to the objectives and proposed model of the study is designed based on TOE framework. The questionnaire is distributed among SOA experts through the large social network, LinkedIn. The questionnaire includes five sections and each section consists of some parts for each factor of the study. The questions of section one are about the respondents demographics and their organizations information, section two covers the technological view of organizations, section three asks questions about the organization aspects, section four includes the questions related to SOA adoption and organization performance, and finally section five asks respondents about their recommendations for leading the organization to be successful in adopting SOA. The questions in sections two, three, and four are prepared in the form of five-point Likert scale evaluation. The answer of these questions arranged as the following: (1) strongly disagree, (2) disagree, (3) neutral, (4) agree, and (5) strongly agree. In the next step, the context of questionnaire will be discussed.

**Section 1:** In this part *Demographic data* is collected by asking six questions related to the global information of organization and the current position of SOA in organization.

**Section 2:** This part consists of questions related to technological aspects of organization and includes part "a", "b," and "c."

(a) **Complexity**

- Q7. Learning to use SOA technologies and associated standards is complicated.
- Q8. Choosing the right standard for our SOA implementations required lots of researching and prototyping.
- Q9. In general SOA is very complex to use.

(b) **Security Concerns**

- Q10. SOA implementations within our organization are supported with secure infrastructure.
- Q11. SOA provides secure services in our organization.
- Q12. Personally, I do not have any concern about the security and privacy of implementing SOA.

(c) **Cost**

- Q13. SOA decreases the investment in new IT project.
- Q14. SOA eliminates the cost of upgrading the legacy system.
- Q15. SOA decreases the cost of system maintenance.

**Section 3:** This part assesses the organizational factors which affect SOA adoption. It consists of five parts, each part evaluates a factor: (a) culture and communication, (b) governance, (c) strategy, (d) business and IT alignment, and (e) ROI.

(a) **Culture and Communication**

- Q16. Paying attention to establish communication between business and IT department is very important.
- Q17. Our employees willing for change towards SOA adoption.
- Q18. Adequate communication of all service orientation project-stakeholders effect on SOA adoption.

(b) **Governance**

- Q19. Our organization has established SOA governance that is fully integrated within IT governance.
- Q20. Definition of organizational responsibilities for managing the service landscape is critical for our organization.
- Q21. Definition of processes for service development and service adaptation is very important in our organization.
- Q22. Definition of service ownerships is vital in SOA adoption.

(c) **Strategy**

- Q23. Our organization's IT strategy support application integration with internal and/or external application services.
- Q24. Our organization's SOA strategy is dependent on business strategy.

(d) **Return on Investment (ROI)**

- Q25. Implementing SOA increased revenue and reduced costs in our organization.
- Q26. Overall, SOA implementation positively increased return on investment.

(e) **Business and IT Alignment**

- Q27. The IT strategy is accurately aligned with the business strategy in our organization.
- Q28. The IT investments are accurately aligned with the business objectives in our organization.
- Q29. The business strategy is effectively supported by the IT strategy in our organization.

**Section 4:** In this section, SOA adoption and the performance of organization are evaluated by asking four questions.

(a) **SOA Adoption**

- Q30. Moving to a Service-Oriented Architecture (SOA) brings many benefits to businesses.
- Q31. Agility for business is delivered by aligning IT infrastructure with business requirements through a well-implemented SOA.

(b) **Organizational Performance**

- Q32. Implementing SOA positively increases the performance of our organization.
- Q33. Adoption of SOA leads our organization to better performance.

**Section 5:** This section includes a multiple answer question to ask experts about their recommendations for being successful in the way to adopt SOA in organization. The answers of this question are adopted based on the five best practices which IBM suggested for developing a successful SOA [19].

**Recommendation**

- Q34. What is your recommendation to be successful in SOA adoption process based on your own experience? (Multiple answers)
- Develop an architecture with a vision for the future.
- Foresee linkages from IT to your business processes.
- Create an organizational culture and skills to support SOA.
- Build a scalable infrastructure.
- Enable operational visibility through governance and service management.
- Other: _____

## 4.5   The Participants of Research

As it is mentioned in the previous chapter, the researcher used self-selecting sampling in this study. So, the link of online questionnaire was posted in the professional groups in LinkedIn such as *Arcitura IT Certified Professionals* and *SOA Professionals Worldwide*. Moreover, to speed up the process, the questionnaire is posted to SOA experts individually through LinkedIn. The researcher selected experts from the members of both professional groups. For achieving the best consequence, before sending the link of the questionnaire through a message the profile of each proficient is checked. SOA experts are selected from the following categories:

- CIO, CTO, Chief Technical Architect, CSO/CISO, VP of IS/IT
- IS Manager, Director, Planner
- IS/IT/Technical Architect
- Other IT Manager in IS Department
- IT Staff

## 4.6   Questionnaire Validity and Reliability and Pilot Study

In this section, the validity and reliability of the questionnaire is examined. For validating the questionnaire, researcher used discriminant and convergent validity while for examining the reliability of the questionnaire Cronbach's alpha is used.

In continuation, the validity and reliability of the questionnaire is examined with 40 data for pilot study.

### 4.6.1  Validity

In preparing a suitable questionnaire, making a plan and consulting with experts are the most significant factors. To collect quantitative data, a questionnaire is the primary need. Preparing questions is based on the goals of the research. After this step, the questionnaires should be distributed among subjects of the study to achieve the required data.

For this study, an adopted questionnaire is designed based on factors selected from Chap. 2, Literature Review. To design the questionnaire several available questionnaires were reviewed. The main questionnaire the researcher used in this study is from the Master's Thesis "Factors affecting adoption of service-oriented architecture (SOA) at an enterprise level" [13]. Moreover, Master Thesis "Factors Influencing the Adoption of Cloud Computing by Small and Medium-Sized" [20] is another source this study used for designing the questionnaire. The researcher also used a paper titled "Critical Success Factors of Service Orientation in Information Systems Engineering" for adopting some parts of the questionnaire [3]. The questionnaire items and sources are provided in Table 4.3.

**Table 4.3** Questionnaire items and sources

| Section | Factor | Role | Question no. | Source |
|---|---|---|---|---|
| Demographic | – | – | Q1, Q2, Q3, Q4, Q5, Q6 | Lennan [13] |
| Technology context | Complexity | Independent | Q7, Q8 | |
| | | | Q9 | Tehrani [20] |
| | Security concerns | Independent | Q10 | Lennan [13] |
| | | | Q11, Q12 | Tehrani [20] |
| | Costs | Independent | Q13, Q14, Q15 | |
| Organizational context | Culture and communication | Independent | Q16, Q18 | Lennan [13] |
| | | | Q17 | Aier et al. [3] |
| | Governance | Independent | Q19 | Lennan [13] |
| | | | Q20, Q21, Q22 | Aier et al. [3] |
| | Strategy | Independent | Q23, Q24 | Lennan [13] |
| | ROI | Independent | Q25, Q26 | Researcher Develop |
| | Business IT alignment | Independent | Q27, Q28, Q29 | Researcher Develop |
| Adoption and performance | SOA adoption | Dependent | Q30, Q31 | Researcher Develop |
| | Organization performance | Dependent | Q32, Q33 | Researcher Develop |
| Recommendation | – | – | Q34 | IBM [19] |

It is notable that some questions are designed by researcher. After creating questionnaire, the validity of the questions is verified with *Dr. Ab Razak Che Hussin* and two other experts, *Dr. Imran Ghani* and *Dr. Amin Saedi*. Dr. Imran is senior lecturer in Faculty of Computing at UTM and service-oriented architecture is one of his fields of study. Dr. Amin is a PhD student of Information System at UTM. The field of his study is about cloud computing adoption. Moreover, the researcher conducted a pilot study for investigating the reliability and validity of the questionnaire, too.

In addition, there are different ways for investigating the validity of a questionnaire. In this study, researcher used discriminant and convergent validity to make evidence for validating the questionnaire of this study.

**Discriminant Validity**

Discriminant validity evaluates the degree of the constructs that are different from each other. According to [21], the minimal value of AVE is 0.5. It is presented the adequate convergent validity. It means that a latent construct is able to explain greater than half of the variance of the questions on standard [22, 23]. As it is clear in Table 4.4, the AVE value for all of the constructs is more than 0.5.

By comparing the relationship between square root of AVE and constructs, the discriminant validity will be measured. A measure of the error free variance for a set of objects is known as AVE. In other words, AVE indicates the totality of variance in the questions computed by the latent construct [22, 24]. Table 4.5 illustrates that discriminant validity between all constructs and AVE square roots are acceptable.

Table 4.6 represents the cross loading output. This table shows the loading of each cell that is greater than other cells in its column and row. The cross loading results estimate the discriminant validity that are acceptable for all constructs and indicators. It can be concluded; the proposed framework has accepted its discriminant validity.

**Table 4.4** AVE square root

| | AVE | AVE square root |
|---|---|---|
| BIA | 0.644965 | 0.803097 |
| Complexity | 0.690348 | 0.830872 |
| Costs | 0.657259 | 0.810715 |
| Culture and communication | 0.768161 | 0.876448 |
| Governance | 0.767387 | 0.854001 |
| Organization performance | 0.858301 | 0.926445 |
| ROI | 0.802801 | 0.895992 |
| SOA adoption | 0.715556 | 0.845905 |
| Security concerns | 0.704647 | 0.839432 |
| Strategy | 0.824441 | 0.907987 |

**Table 4.5** Inter-correlations of the latent variables and AVE square roots

|      | BIA | Compx | Cst | Cul | Gov | OP | ROI | SOA | Sec | Str |
|------|-----|-------|-----|-----|-----|----|-----|-----|-----|-----|
| BIA | **0.803097** | | | | | | | | | |
| Compx | −0.581705 | **0.830872** | | | | | | | | |
| Cst | −0.115180 | 0.192262 | **0.810715** | | | | | | | |
| Cul | 0.730976 | −0.388590 | −0.409906 | **0.876448** | | | | | | |
| Gov | 0.773021 | −0.504991 | −0.149643 | 0.462978 | **0.854001** | | | | | |
| OP | 0.584725 | −0.587577 | −0.269833 | 0.517567 | 0.535866 | **0.926445** | | | | |
| ROI | 0.675227 | −0.357958 | −0.226868 | 0.708188 | 0.436204 | 0.566369 | **0.895992** | | | |
| SOA | 0.717202 | −0.601689 | −0.380522 | 0.781992 | 0.744074 | 0.766448 | 0.679753 | **0.845905** | | |
| Sec | −0.556062 | 0.388326 | 0.191996 | −0.611078 | −0.657254 | −0.552846 | −0.429475 | −0.691444 | **0.839432** | |
| Str | 0.757631 | −0.526189 | −0.271529 | 0.634614 | 0.708666 | 0.621213 | 0.555052 | 0.811947 | −0.670532 | **0.907987** |

*Compx* Complexity, *Sec* Security, *Cul* Culture and Communication, *Gov* Governance, *BIA* Business and IT Alignment, *ROI* Return on Investment, *SOA* SOA adoption, *OP* Organizational Performance

**Table 4.6** The cross loading output

| | BIA | COMPX | CST | CUL | GOV | OP | ROI | SOA | SEC | STR |
|---|---|---|---|---|---|---|---|---|---|---|
| BIA1 | **0.795203** | -0.515057 | -0.070138 | 0.549126 | 0.606662 | 0.495609 | 0.495044 | 0.634122 | -0.444787 | 0.609637 |
| BIA2 | **0.725225** | -0.445665 | 0.069248 | 0.547369 | 0.473580 | 0.371470 | 0.488314 | 0.548826 | -0.384850 | 0.413948 |
| BIA3 | **0.874780** | -0.452502 | -0.257789 | 0.667941 | 0.742115 | 0.546717 | 0.617116 | 0.777349 | -0.490903 | 0.743568 |
| COMPX1 | -0.550638 | **0.840346** | 0.078663 | -0.379074 | -0.471169 | -0.552510 | -0.404782 | -0.528488 | 0.346737 | -0.461859 |
| COMPX2 | -0.407830 | **0.765380** | 0.198234 | -0.249632 | -0.268194 | -0.463114 | -0.179606 | -0.432798 | 0.212911 | -0.387086 |
| COMPX3 | -0.486142 | **0.882640** | 0.210576 | -0.329667 | -0.495617 | -0.449227 | -0.289402 | -0.530877 | 0.398344 | -0.457592 |
| CST1 | 0.198123 | 0.073391 | **0.706120** | -0.018219 | 0.035231 | -0.085996 | 0.016623 | -0.062392 | -0.101208 | 0.069583 |
| CST2 | -0.025955 | -0.015354 | **0.773863** | -0.323438 | 0.039443 | -0.160831 | -0.142660 | -0.250849 | 0.105872 | -0.219408 |
| CST3 | -0.211702 | 0.288759 | **0.935169** | -0.420285 | -0.255305 | -0.297151 | -0.258683 | -0.413565 | 0.157902 | -0.287532 |
| CUL1 | 0.724133 | -0.348378 | -0.399993 | **0.924806** | 0.475029 | 0.485578 | 0.707353 | 0.707360 | -0.476943 | 0.609729 |
| CUL2 | 0.643228 | -0.396102 | -0.361741 | **0.850806** | 0.383029 | 0.531784 | 0.664991 | 0.711165 | -0.595466 | 0.593768 |
| CUL3 | 0.564873 | -0.269549 | -0.310715 | **0.851671** | 0.354078 | 0.330126 | 0.474141 | 0.631707 | -0.447167 | 0.454317 |
| GOV1 | 0.754022 | -0.551599 | -0.178335 | 0.454929 | **0.938299** | 0.482960 | 0.430258 | 0.717348 | -0.580245 | 0.656143 |
| GOV2 | 0.543451 | -0.333110 | -0.091204 | 0.406297 | **0.803231** | 0.407964 | 0.369018 | 0.531471 | -0.517348 | 0.524621 |
| GOV3 | 0.766900 | -0.522185 | -0.157972 | 0.391317 | **0.860733** | 0.608796 | 0.413113 | 0.709462 | -0.558950 | 0.693291 |
| GOV4 | 0.606283 | -0.325115 | -0.081760 | 0.370057 | **0.896153** | 0.354124 | 0.307942 | 0.622398 | -0.574746 | 0.588367 |
| OP1 | 0.630344 | -0.568467 | -0.228329 | 0.497842 | 0.497702 | **0.928714** | 0.559096 | 0.720546 | -0.488619 | 0.643909 |
| OP2 | 0.471772 | -0.519562 | -0.272233 | 0.460647 | 0.495224 | **0.924171** | 0.489341 | 0.699450 | -0.501916 | 0.505097 |
| ROI1 | 0.586681 | -0.247076 | -0.172488 | 0.666746 | 0.302474 | 0.437056 | **0.881575** | 0.567335 | -0.412982 | 0.448056 |
| ROI2 | 0.614902 | -0.385925 | -0.230509 | 0.607354 | 0.469065 | 0.570129 | **0.910178** | 0.646537 | -0.246216 | 0.541408 |
| SEC1 | -0.538198 | 0.356284 | 0.165424 | -0.586093 | -0.628533 | -0.491573 | -0.326848 | -0.632038 | **0.923020** | -0.611145 |
| SEC2 | -0.318830 | 0.321366 | -0.050515 | -0.361565 | -0.350449 | -0.348431 | -0.292728 | -0.422444 | **0.767489** | -0.327873 |
| SEC3 | -0.519667 | 0.327053 | 0.185137 | -0.505123 | -0.602109 | -0.504169 | -0.307137 | -0.608136 | **0.859640** | -0.547208 |
| SOA1 | 0.635792 | -0.500394 | -0.417778 | 0.583353 | 0.604522 | 0.684667 | 0.520265 | **0.848359** | -0.540347 | 0.730316 |

| SOA2 | 0.763263 | −0.517663 | −0.224736 | 0.740554 | 0.654626 | 0.611611 | 0.630403 | **0.843441** | −0.580947 | 0.642850 |
|------|----------|-----------|-----------|----------|----------|----------|----------|----------|-----------|----------|
| STR1 | 0.689997 | −0.419783 | −0.234430 | 0.549402 | 0.663107 | 0.467530 | 0.467615 | 0.705378 | −0.560474 | **0.899857** |
| STR2 | 0.678729 | −0.531392 | −0.257627 | 0.601257 | 0.625828 | 0.653160 | 0.537745 | 0.767135 | −0.529898 | **0.916047** |

*Compx* Complexity, *Sec* Security, *Cul* Culture and Communication, *Gov* Governance, *Cst* Cost, *Str* Strategy, *BIA* Business and IT Alignment, *ROI* Return on Investment, *SOA* SOA adoption, *OP* Organizational Performance

**Table 4.7** Factor loadings

|  | Factor loadings |
| --- | --- |
| C0MPX1 | 0.840 |
| COMPX2 | 0.765 |
| COIVIPX3 | 0.883 |
| SEC1 | 0.917 |
| SEC2 | 0.744 |
| SEC3 | 0.849 |
| CST1 | 0.706 |
| CST2 | 0.774 |
| CST3 | 0.935 |
| GOV1 | 0.938 |
| GOV2 | 0.803 |
| GOV3 | 0.861 |
| GOV4 | 0.896 |
| STR1 | 0.900 |
| STK2 | 0.916 |
| CUL1 | 0.925 |
| CUL2 | 0.851 |
| CUL3 | 0.852 |
| BIA1 | 0.796 |
| BIA2 | 0.728 |
| BIA3 | 0.878 |
| ROI1 | 0.882 |
| ROI2 | 0.910 |
| SOA1 | 0.848 |
| SOA2 | 0.844 |
| OP1 | 0.929 |
| OP2 | 0.924 |

*Compx* Complexity, *Sec* Security, *Cul* Culture and Communication, *Gov* Governance, *Cst* Cost, *Str* Strategy, *BIA* Business and IT Alignment, *ROI* Return on Investment, *SOA* SOA adoption, *OP* Organizational Performance

## Convergent Validity

At last, factor loadings obtained the significant level. Table 4.7 shows that the value of all items in proposed framework are more than 0.7 which illustrated that all constructs meet the adequate acceptance and achieved the convergent validity.

**Table 4.8** Cronbach's alpha

|  | Cronbach's alpha |
|---|---|
| BIA | 0.724383 |
| Complexity | 0.774867 |
| Costs | 0.766402 |
| Culture and communication | 0.848392 |
| Governance | 0.898522 |
| Organization performance | 0.834943 |
| ROI | 0.755337 |
| SOA adoption | 0.602497 |
| Security | 0.790590 |
| Strategy | 0.787415 |

## *4.6.2 Reliability*

As it is mentioned in Chap. 3, the reliability of questionnaire is measured by Cronbach's alpha. The coefficient alpha is the most accepted measure of reliability which determines how a series of questions explain a single construct. [25] stated in a study that the minimum accepted level of Cronbach's alpha for an exploratory study is 0.60 (As cited in [13, 26]). Table 4.8 shows the value of constructs in the pilot study of this research. As it is clear, all constructs have appropriate and acceptable value of coefficient alpha.

## 4.7 Chapter Summary

According to the findings and the relationship among influential factors, SOA adoption and organizational performance, this chapter proposed a framework based on TOE framework. Besides, researcher had described about the hypothesis of the framework. Moreover, the five sections of the questionnaire are discussed. First section is about demographic data, second section is about technology aspect of organization, third part is belonged to organization aspects, fourth section is about the SOA adoption and organizational performance, and section five collects the experts' recommendations for being successful in adopting SOA and increasing the performance of organizations. Finally, the validity and reliability of the questionnaire were measured by AVE Square root and Cronbach Alpha and then the measurement model assessment was provided.

## References

1. Rogers, E. M. (1983). Diffusion of innovations (3rd ed.). New York, NY: The Free Press
2. J.H. Lee, H.-J. Shim, K.K. Kim, Critical success factors in SOA implementation: an exploratory study. Inf. Syst. Manag. **27**, 123–145 (2010)

3. S. Aier, T. Bucher, R. Winter, Critical success factors of service orientation in information systems engineering. Bus. Inf. Syst. Eng. **3**, 77–88 (2011)
4. J. Antikainen, S. Pekkola, Factors influencing the alignment of SOA development with business objectives, in 17th European Conference on Information Systems (ECIS), 2009, pp. 2579–2590 (2009)
5. B.H. Reich, I. Benbasat, Factors that influence the social dimension of alignment between business and information technology objectives. MIS Quarterly. **24**(1), 81–113 (2000)
6. Strassmann, Paul A. (1989), What is Alignment? Alignment is The Delivery of the Required Results, The Squandered Computer, Cutter IT Journal, August 1998
7. W. Vegter, Critical success factors for a SOA implementation, in *11th Twente Student Conference on IT*, Enschede, June 29th, 2009
8. C. Hu, *Computing and Intelligent Systems* (Springer, London, 2011), pp. 81–87
9. A. Seth, A.R. Singla, H. Aggarwal, *Contemporary Computing* (Springer, Berlin, 2012), pp. 164–175
10. A.P. Ciganek, M.N. Haines, W.D. Haseman, Service-oriented architecture adoption: key factors and approaches. J. Inf. Technol. Manag. **203**, 42–54 (2009)
11. Jeff, Service Oriented Enterprise: Budgeting for SOA, 2007. [Online]. Available: http://schneider.blogspot.com/2007/09/budgeting-for-soa.html. [Accessed: 30-Jan-2010].
12. M. Galinium, N. Shahbaz, Factors affecting success in migration of legacy systems to service-oriented architecture (SOA), Master Thesis, Department of Informatics, Lund University, Sweden, Submitted, 2009
13. E.M. Lennan, Factors affecting adoption of service-oriented architecture (SOA) at an enterprise level, Department of Information Systems, University of Cape Town, 2011
14. T. Erl, *Service-Oriented Architecture Concept, Technology and Design* (Pearson Inc., Upper Saddle River, 2005)
15. M. Findikoglu, *Web Services Adoption Process: A Roadmap to Manage Organizational Change Available at SSRN 1873723*, 2011
16. B. Roulstone, J.J. Phillips, *ROI for Technology Projects* (Routledge, London, 2012)
17. J. DiMare, *Service Oriented Architecture: A Practical Guide to Measuring Return on That Investment* (IBM Global Business Services, 2006)
18. G.A. Gelade, M. Ivery, The impact of human resource management and work climate on organizational performance. Pers. Psychol. **56**, 383–404 (2003)
19. IBM, Five best practices for deploying a successful service-oriented architecture, IBM Global Services, 2008. [Online]. Available: http://www-935.ibm.com/services/us/its/pdf/wp_five-best-practices-for-deploying-successfulsoa.pdf. [Accessed: 18-Apr-2013]
20. S.R. Tehrani, *Factors Influencing the Adoption of Cloud Computing by Small and Medium-Sized Enterprises (SMEs)*, 2013
21. W. W. Chin, The Partial Least Squares Approach to Structural Equation Modeling, in Modern methods for business research, 295–336 (1998)
22. N. Mohammadhossein, A study on the CRM customer benefits towards customer satisfaction, Faculty of Computing, Universiti Teknologi Malaysia, 2013
23. S. Akter, J. D'Ambra, P. Ray, Service quality of mHealth platforms: development and validation of a hierarchical model using PLS. Electron. Mark. **20**, 209–227 (2010)
24. B.H. Wixom, P.A. Todd, A theoretical integration of user satisfaction and technology acceptance. Inf. Syst. Res. **16**, 85–102 (2005)
25. J. C. Nunnally, I. H. Bernstein. Psychometric theory. Vol. 226. New York: McGraw-Hill, 1967
26. E. Ngai, T. Cheng, S. Ho, Critical success factors of web-based supply-chain management systems: an exploratory study. Prod. Plan. Control **15**, 622–630 (2004)

# Chapter 5
# Analyzing of Service-Oriented Architecture (SOA) Experts Responses by SmartPLS Version 2

## 5.1 Introduction

In this chapter, researcher deals with data collection and data analyzing which are collected from the questionnaires that were distributed among experts through LinkedIn. In this section, analyzing data and testing the hypothesis are discussed in detail. It is worth mentioning that for data examining and analyzing researcher uses SmartPLS software. At the first step of this chapter, demographic information will be analyzed and then the effect of significant factors on SOA adoption and organizational performance will be examined. Finally, at the end of this chapter some recommendations for being successful in adopting SOA and improving the performance of organizations based on the information extracted from the last question of the questionnaire will be developed.

## 5.2 Questionnaire Development

Based on the objectives of the study, after doing pilot study and a little change the questionnaire was sent to the rest of experts. As it described before, a total number of 369 questionnaires were sent to experts through LinkedIn and had received 117 responses. After checking the feedbacks it is found that only 104 questionnaires are useful for analyzing. Questionnaire consists of five sections as following:

- **First section:** Demographic data
- **Second section:** Technological aspects of organization (complexity, security, and costs)

- **Third section:** Organizational context (Culture and communication, ROI, Business and IT alignment, Governance, and Strategy)
- **Fourth section:** SOA adoption and organizational performance
- **Fifth section:** Recommendations

## 5.3    Data Analysis

### 5.3.1    Demographic Information

In this section, demographic data are explained in the view of graphs and each graph is described briefly as follows.

**Job Position of the Respondents**

Figure 5.1 illustrates that more than half of the respondents came from IS/IT/technical architects group (59%). The next largest group that participate in this research were CIO, CTO, Chief Technical Architects, CSO/CISO, VP of IS/IT (17%) and IS managers, planners, and directors (13%). The rest of respondents are IT staff (3%) and other IT managers in IS department (4%).

**Industry**

Figure 5.2 indicates that most of the participants came from consulting and business services industry (26.92%). In addition, 18.27% of participants were from the Telecommunications/ISP section, 15.38% were from IT vendors, 9.62% from

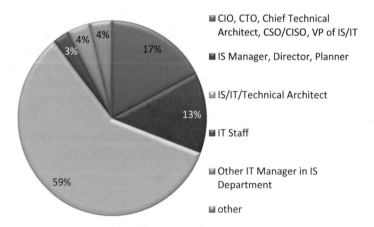

**Fig. 5.1** Respondents position in organizations ($N = 104$)

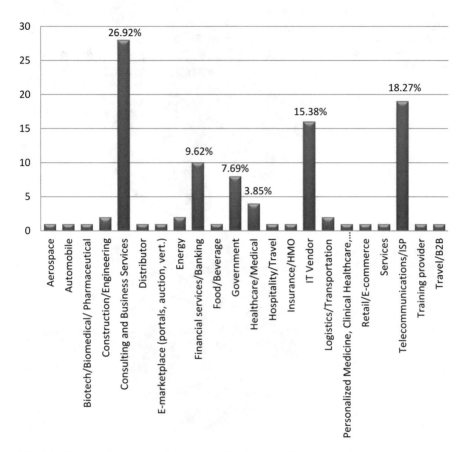

**Fig. 5.2** The industry of the participants' organization (*N* = 104)

financial services/banking, and 7.69% were from governmental organizations. The rest of the participants from different sectors are represented by less than 5%.

**Number of Employees**

Most of the participants are working in large and very large companies. About 38% are from companies with 500–5000 employees and 36% are from companies with more than 5000 employees. Fifteen percent of the respondents are working in medium-sized companies with 50–499 employees. Small companies are represented by 12% of the participants (Fig. 5.3).

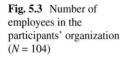

**Fig. 5.3** Number of employees in the participants' organization ($N$ = 104)

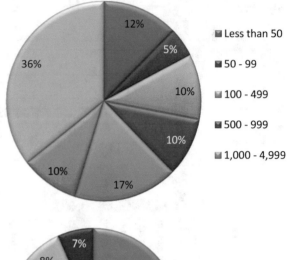

**Fig. 5.4** Success of SOA project ($N$ = 104)

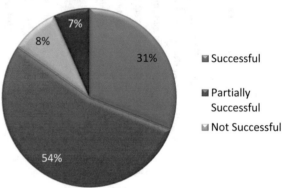

## Success of SOA Project

More than half of participants (54%) illustrated that their organization is partially successful in developing SOA in their organization while 31% of respondents claimed that their organization is completely successful in implementing SOA. Only 8% of respondents indicated that their SOA implementation is not successful and 7% of participants declared it is too early to tell about their failure or success of implementing SOA in their organizations (Fig. 5.4).

## Stage of SOA Adoption

A large number of participants (62%) indicated that their SOA deployments are in production. Thirty three percent deployed in production at enterprise level and 29% deployed in production for use in multiple departments. Fourteen percent of respondents pointed out that their SOA implementation are in the pilot stage while 8% of

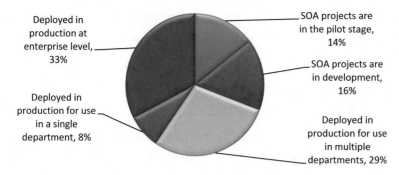

**Fig. 5.5**  Stage of SOA adoption (*N* = 104)

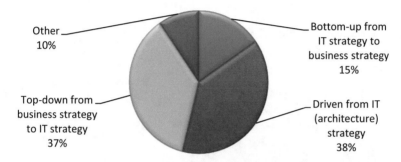

**Fig. 5.6**  Approach of the organization for starting SOA (*N* = 104)

them illustrated they are in production for use in a single department. The rest of respondents (16%) pointed that their SOA projects are in development (Fig. 5.5).

**SOA Initiative Approach**

Figure 5.6 shows the approach of respondents to SOA initiative. As it is clear, 38% of respondents indicated that their SOA is driven from IT Strategy and 37% of them illustrated they used top-down approach which is driven from business strategy to IT strategy. Only 15% of participants used bottom-up approach which is driven from IT strategy to business strategy. The rest of the participants (10%) used other approach to SOA initiative.

**Participants' Nationality**

Figure 5.7 represents the distribution of participants of this study. As it is clear in this figure, most of the respondents (39.40%) were from Brazil (22.22%) and the USA (17.17%) while about 17.15% of participants were from India, Netherlands,

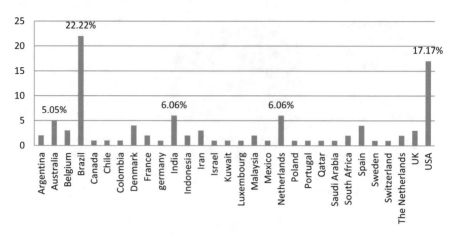

**Fig. 5.7** Participants' nationality ($N = 99$)

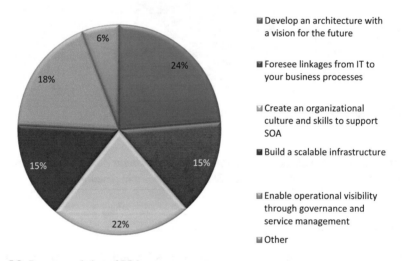

**Fig. 5.8** Recommendation of SOA experts

and Australia. There were some people from other countries as it is shown in the following figure.

## Recommendation of SOA Experts

Based on the third objective of this study, an adopted question is designed to find out the recommendation of SOA professionals for being successful in adopting SOA based on their own experience. As it is clear in Fig. 5.8, most experts (46%) agreed with developing an architecture with a vision for the future and creating an organizational culture and skills to support SOA. Eighteen percent of experts (18%) believed that enabling operational visibility through governance and service management helps organizations to adopt SOA successfully while 30% of professionals accepted that building a scalable infrastructure (15%) and foreseeing linkages from IT to the business processes (15%) assisted organizations to adopt SOA successfully. Six percent of professionals write their own recommendations in the "Other" field which will be described in the last chapter of this study.

As it is illustrated in Fig. 5.8, there is a field in questionnaire, *Other*, for experts to write their recommendations. The following table shows the experts' suggestions towards success of SOA adoption (Table 5.1).

## 5.3.2   Hypothesis Testing

As it is mentioned before, for approving the structural framework and for testing the hypothesis associations in this study researcher used PLS technique. Due to PLS capability to design latent constructs controlled by conditions for small to medium sample size, it has recently become very popular among researchers [1].

**Table 5.1**  Experts' recommendations

| | |
|---|---|
| Expert 1 | Determine the criticality of the system that need to be migrated/integrated to SOA-based system |
| Expert 2 | Start with a little scope, increasing progressively |
| Expert 3 | Define a roadmap, align against capabilities and business processes, align maturity and goals to roadmap |
| Expert 4 | Establish a SOA roadmap at the beginning of the adoption |
| Expert 5 | SOA is a strategic business view, TOP DOWN is the best and the only valid strategy |
| Expert 6 | Support infrastructure maintenance actively |
| Expert 7 | Focus on the seven strategic goals of SOA |
| Expert 8 | Do not try to build yourself a SOA, but give a solution room to grow into a SOA. Use rules and principles and technology in the correct way, use common sense and cherish your vision and what is beyond |
| Expert 9 | Start small and show quick ROI before going for enterprise wide |
| Expert 10 | Training on all IT levels, from managers to developers, and even system users if necessary |
| Expert 11 | Business and IT should work together not only aligned |
| Expert 12 | Be ready to change quickly |
| Expert 13 | Use business terms rather than technology or applications terms like SAP, Oracle. Business should think as services not applications like SharePoint, SAP, etc. |
| Expert 14 | Start small and grow up SOA adoption with small steps |
| Expert 15 | Be generic in service layer to handle any type of client system request |
| Expert 16 | Start small, but visible. Address current business issues |
| Expert 17 | Make sure your customer has high level backing for the implementation |
| Expert 18 | Make business owners service owners and give them incentives to cooperate across business units |

In the previous chapter (Chap. 4), the reliability and validity of the questionnaire was tested by evaluating and analyzing the questions. The consequence of analyzing the questionnaire was described completely. In this chapter, researcher examines the framework and the paths between constructs to find the significance of hypothesis in the framework. The relationships among the different latent constructs were analyzed to achieve this goal. Therefore, 104 data collected from questionnaire import to SmartPLS to discover the importance and significance of the paths. Figure 5.8 presents the framework of this study.

A prominent method for statistical issues is bootstrapping which is used to estimate the changeability of statistical test. Bootstrap is a non-parametric technique for evaluating the function, discovering the errors, and estimating the significance of single regressors [2]. Bootstrapping measures $t$ statistics (or $t$-value) that correlates with different model paths (inner and outer). Moreover, the probability value or $p$-value is used to present the evidence of accepting hypotheses and to measure the statistical significance for testing the hypotheses. The maximum acceptance value for $p$-value is 0.05.

Standard beta coefficient (std. $\beta$) is used to examining hypothesis. Besides, for generalizing the coefficient of relationship, the multiple correlation coefficient is utilized. To assessing the quality of the divination of the dependent variable, the correlation coefficient is used in various regression analyses. It correlates to the squared relationship between the real and anticipated values of the dependent variable [3].

Figure 5.9 presents the framework of this study with calculated $t$-values, path coefficient $\beta$ related to the connection of each factors and SOA adoption and the path loadings of all factors and $R^2$ for SOA adoption and organizational performance.

The hypothesis results are depicted in Table 5.2. This table shows the consequences of the relationships which are proposed for this study, also.

The above table (Table 5.2) includes the $t$-value and the coefficient for the relationship between each factor and SOA adoption and the effect of SOA adoption on organizational performance. As it is clear in this table complexity, security, and costs negatively effect on SOA adoption while governance, strategy, culture and communication, Business and IT alignment, and ROI positively effect on SOA adoption. Moreover, SOA adoption itself has a positive impact on organizational performance.

As it is mentioned before, squared multiple correlation coefficient is used to measure the correlation between independent and dependent variables. Figure 5.7 shows that the correlation between factors and SOA adoption is 0.86 and the correlation between SOA adoption and organizational performance is 0.54. It means that factors 86% influence on SOA adoption while SOA adoption impact 54% on the performance of organizations based on data collected in this study. Table 5.3 shows the squared multiple correlation for the framework of this study.

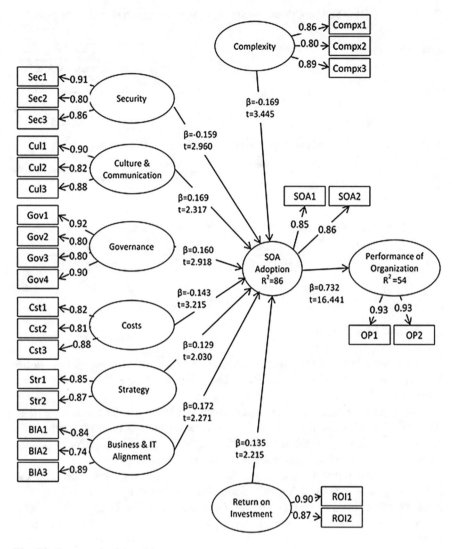

**Fig. 5.9** Framework of the study

**Table 5.2** The result of hypothesis

|    | Hypothesized path | β | t-value | Result |
|----|-------------------|---|---------|--------|
| H1 | Complexity → SOA adoption | −0.169 | 3.367 | Accept |
| H2 | Security → SOA adoption | −0.159 | 2.827 | Accept |
| H3 | Costs → SOA adoption | −0.143 | 3.108 | Accept |
| H4 | Governance → SOA adoption | 0.160 | 2.620 | Accept |
| H5 | Strategy → SOA adoption | 0.129 | 2 | Accept |
| H6 | Culture and communication → SOA adoption | 0.169 | 2.267 | Accept |
| H7 | Business and IT alignment → SOA adoption | 0.172 | 2.212 | Accept |
| H8 | ROI → SOA adoption | 0.135 | 2.143 | Accept |
| H9 | SOA adoption → organization performance | 0.732 | 16.339 | Accept |

**Table 5.3** $R$ square

| | $R$ square |
|---|---|
| Organization performance | 0.535200 |
| SOA adoption | 0.860012 |

**Table 5.4** Analysis of the relationship between complexity and SOA adoption

| | Item | Coefficient | $t$-value | $p$-value |
|---|---|---|---|---|
| H1 | Complexity | −0.169 | 3.367 | $1.06606*10^{-3}$ |

**Table 5.5** Analysis of the relationship between security and SOA adoption

| | Item | Coefficient | $t$-value | $p$-value |
|---|---|---|---|---|
| H2 | Security | −0.159 | 2.827 | $5.63636*10^{-3}$ |

**Analysis of First Hypothesis**

The First hypothesis of this study is: *"The Complexity of SOA Technology negatively influence on the adoption of SOA."* The following table (Table 5.4) shows the result of first hypothesis.

As it is seen in the table, the amount of $p$-value is 0.001066061 that is lesser than 0.05. Besides, the result of this analysis shows that complexity influences negatively on SOA adoption, since the sign of path coefficient value is negative (−0.169). Therefore, it can be concluded this hypothesis is acceptable.

It can be deliberated that the complexity of SOA technology would decrease the profit of adopting SOA in organizations. This outcome is in line with earlier researches that stated the complexity of SOA technology is a difficulty that related to applying and understanding of a technology and negatively effect on the adoption level [4].

**Analysis of Second Hypothesis**

The second proposed hypothesis of this study is: *"The security concerns negatively influence on SOA adoption."* Table 5.5 presents the result of this hypothesis.

Security concern is one of the significant factors that cited many times in previous studies. Table 5.5 illustrated that security concerns have affected on SOA adoption negatively as it is predicted by H2, forasmuch as the coefficient value is negative. Whereas the $p$-value obtained is less than 0.05 for this factor, so this hypothesis is accepted. It is concluded that security concerns prevent organization to adopt SOA. This hypothesis is in line with previous studies that demonstrated security as a common barrier for adopting SOA [5, 6].

**Table 5.6**  Analysis of the relation between costs and adopting SOA

|      | Item  | Coefficient | $t$-value | $p$-value |
|------|-------|-------------|-----------|-----------|
| H6   | Costs | −0.143      | 3.108     | $2.42934*10^{-3}$ |

**Table 5.7**  Analysis of the relationship between governance and SOA adoption

|      | Item       | Coefficient | $t$-value | $p$-value |
|------|------------|-------------|-----------|-----------|
| H4   | Governance | 0.160       | 2.620     | $1.01086*10^{-2}$ |

**Analysis of Third Hypothesis**

As it is discussed in the previous chapter, the third hypothesis is: *"Costs negatively influence on SOA adoption."* Table 5.6 shows the result of testing this hypothesis.

The result of this hypothesis illustrated that cost negatively effect on adopting SOA, since path coefficient value is minus (−0.143) and $t$-value is 3.108. Therefore, $p$-value becomes smaller than 0.05. These results show that there is a significant relationship between costs and SOA adoption. Thus, this hypothesis is acceptable and high costs prevent organization to adopt SOA. This consequence is in line with the findings of Lennan [7] which clarified that the less likely to adopt SOA is the result of the high costs of implementing SOA.

**Analysis of Fourth Hypothesis**

As it is identified before, fourth hypothesis is: *"SOA Governance positively effects on adopting SOA in organizations."* The consequence of fourth hypothesis is shown in Table 5.7. From the results obtained by this hypothesis path coefficient is 0.160 that is depicted this hypothesis positively effect SOA adoption. Moreover, $t$-value is 2.620 and $p$-value is smaller than 0.05. It could be concluded that this hypothesis is acceptable. Mac Lennan [7] has expressed the same result and illustrated that one of the most important features of SOA adoption process is SOA governance. Lee [8] admitted governance as a critical success factor and stated that one of the most important factors for organizations to being successful in adopting SOA is governance.

**Analysis of Fifth Hypothesis**

As it is mentioned before, the fifth hypothesis is: *"SOA Strategy positively influence on SOA implementation."* The result of this hypothesis is presented in Table 5.8. The path coefficient 0.160 and $t$-value 2.620 illustrated that this hypothesis positively effects on SOA adoption. Whereas $p$-value is less than 0.05 therefore it could be concluded that this hypothesis is acceptable.

**Table 5.8** Analysis of the relation between SOA strategy and adopting SOA

|    | Item | Coefficient | $t$-value | $p$-value |
|----|------|-------------|-----------|-----------|
| H5 | Strategy | 0.129 | 2 | $4.81074*10^{-2}$ |

**Table 5.9** Analysis of the relationship between culture and communication and SOA adoption

|    | Item | Coefficient | $t$-value | $p$-value |
|----|------|-------------|-----------|-----------|
| H3 | Culture and communication | 0.169 | 2.267 | $2.54592*10^{-2}$ |

**Table 5.10** Analysis of the relationship between business and IT alignment and SOA adoption

|    | Item | Coefficient | $t$-value | $p$-value |
|----|------|-------------|-----------|-----------|
| H7 | Business and IT alignment | 0.172 | 2.212 | $2.91542*10^{-2}$ |

The result of this hypothesis is in line with a research of Yoon and Carter [9] which stated that organizational strategy is critical success factor in implementing a successful SOA. Lee et al. [8] in a similar study expressed that strategy is a potential factor influence on adopting SOA in organizations.

## Analysis of Sixth Hypothesis

The sixth hypothesis of this study is: *"Culture and communication positively influence on adopting SOA."* Table 5.9 presents the result of this analysis.

The design of traditional IS is completely changed with service orientated architecture essentially. So, it is vital for organizations to foster the culture of employees for willing to such change [10]. Table 5.9 shows that culture and communication positively effect on SOA adoption since the sign of $\beta$ is positive. Since $t$-value is equal to 2.267 and $p$-value is more than 0.05, so this hypothesis is accepted. Besides, this hypothesis is in line with the earlier studies that illustrated fostering culture and communication is a critical success factor for organizations while adopting SOA [8, 10].

## Analysis of Seventh Hypothesis

As it is expressed in Chap. 4, seventh hypothesis is: *"Business and IT alignment positively influence on SOA adoption."* Business and IT alignment is considered as a factor which positively influence on SOA adoption. The results of analyzing this hypothesis are provided in Table 5.10. The results show significant relationship between SOA adoption and business and IT alignment. As it defined in Table 5.10, path coefficient is positive (0.172) and $t$-value is 2.212. It could be concluded that business and IT alignment has strong relationship with SOA adoption through these outcomes. Whereas, $p$-value is less than 0.05 this hypothesis is acceptable.

**Table 5.11** Analysis of the relationship between ROI and SOA adoption

|     | Item                  | Coefficient | $t$-value | $p$-value                |
| --- | --------------------- | ----------- | --------- | ------------------------ |
| H8  | Return on investment  | 0.135       | 2.143     | $3.44452*10^{-2}$        |

**Table 5.12** Analysis of the relationship between SOA adoption and performance of organization

|     | Item                                              | Coefficient | $t$-value | $p$-value          |
| --- | ------------------------------------------------- | ----------- | --------- | ------------------ |
| H8  | SOA adoption $\rightarrow$ organization performance | 0.732       | 16.339    | $1.74898E - 30$    |

**Analysis of Eighth Hypothesis**

The eighth hypothesis that this study should support is: *"Return on Investment positively effects on SOA adoption."* The consequences of this claim are shown in Table 5.11. In many cases, management views all the investments in the same way. ROI is crucial to success the technology development and IT function. It has strong and positive correlation with SOA adoption based on the findings of this study. It is resulted from path coefficient in Table 5.11 ($\beta = 0.172$). As it is mentioned before, for accepting a hypothesis $p$-value should be smaller than 0.05, so it can be concluded that this hypothesis is acceptable.

**Analysis of Ninth Hypothesis**

As it is predicted in Chap. 4, the ninth hypothesis is: *"SOA adoption positively influence on the performance of organization."* The performance of organization includes the real output and consequences of an enterprise as surveyed against its contracted goals or purposes. Table 5.12 represents the result of analyzing the effect of SOA adoption on the performance of organization. This analysis shows that SOA adoption has a strong positive relationship with organizational performance due to the value of path coefficient which is 0.732 and $t$-value which is 16.339. The consequence of this test expressed the acceptance of hypothesis since the obtained $p$-value is less than 0.05.

## 5.4   Conclusion Analysis and Framework Finalization

As it is mentioned before, the objective of this research is to identify the significant factors which influence on SOA adoption in organization and to propose a framework to evaluate the relationship between these factors, SOA adoption and organization performance. The findings of this study chip in the previous studies by identifying the significant factors that affect SOA adoption. These factors are: complexity, security concerns, costs, governance, strategy, culture and communication,

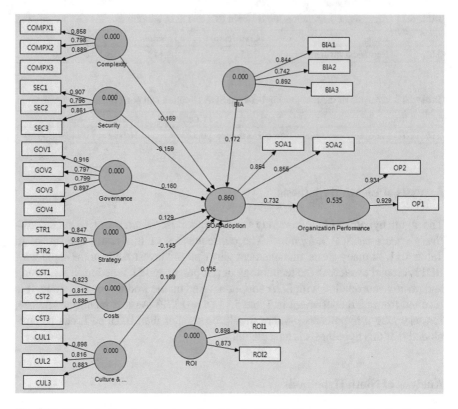

**Fig. 5.10** SOA adoption framework produced by SmartPLS. *Compx* Complexity, *Sec* Security, *Cul* Culture and Communication, *Gov* Governance, *Cst* Cost, *Str* Strategy, *BIA* Business and IT Alignment, *ROI* Return on Investment, *SOA* SOA adoption, *OP* Organizational Performance

business and IT alignment, ROI. Figure 5.10 represents the proposed framework of this study which is produced by SmartPLS.

In order to emphasize the importance of these factors in implementing SOA in organizations, this study presented a framework which concentrates on the association of these factors to implement a successful SOA in organizations. Data was collected from SOA professionals through LinkedIn who have adequate knowledge and experience about SOA in organizations. The results show that three factors negatively effect on SOA adoption, namely complexity, security, and costs while other factors positively influence SOA adoption that are: governance, strategy, culture and communication, business and IT alignment, and ROI. After analyzing the findings of this study, the following framework resulted finally (Fig. 5.11). In this figure, $R^2$ represents the strength of the relationships between factors, SOA adoption and organizational performance. It means these key factors have 86% effect on SOA adoption while SOA adoption has an impact of 54% on organizational performance.

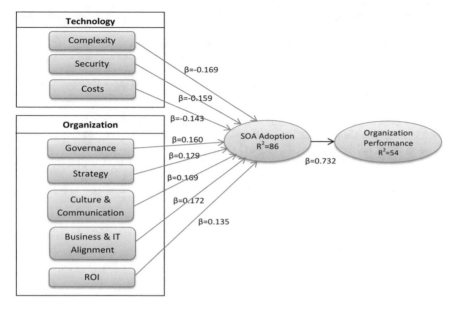

**Fig. 5.11** Finalized framework of the study

## 5.5 Chapter Summary

The final findings and results are covered by this chapter. According to the proposed framework in Chap. 4, an adopted questionnaire was distributed among SOA professionals via LinkedIn. The main goals were to discover potential factors which may influence on SOA adoption. Researcher used SmartPLS to analyze the findings of the study in order to determine the association and the strength of the connections between factors, SOA adoption and organizational performance.

Moreover, the researcher examined the impact of key factors on SOA adoption and organizational performance from path coefficient and *T*-test. The consequences of these analyses represented that three factors negatively affect SOA adoption (complexity, security concerns, and costs) while other factors influence on SOA adoption positively (governance, strategy, culture and communication, business and IT alignment, and ROI).

## References

1. M. Limayem, M. Khalifa, W.W. Chin, CASE tools usage and impact on system development performance. J. Organ. Comput. Electron. Commer. **14**, 153–174 (2004)
2. P.C. Austin, J.V. Tu, Bootstrap methods for developing predictive models. Am. Stat. **58**, 131–137 (2004)

3. H. Abdi, *Multiple Correlation Coefficient* (The University of Texas at Dallas, Richardson, 2007)
4. C. Hu, *Computing and Intelligent Systems* (Springer, Berlin, 2011), pp. 81–87
5. T. Erl, *Service-Oriented Architecture Concept, Technology and Design* (Pearson Inc., Upper Saddle River, 2005)
6. N. Basias, M. Themistocleous, V. Morabito, SOA adoption in e-banking. J. Enterp. Inf. Manag. **26**, 719–739 (2013)
7. E.M. Lennan, Factors affecting adoption of service-oriented architecture (SOA) at an enterprise level, Department of Information Systems, University of Cape Town, 2011
8. J.H. Lee, H.-J. Shim, K.K. Kim, Critical success factors in SOA implementation: an exploratory study. Inf. Syst. Manag. **27**, 123–145 (2010)
9. T. Yoon, P. E. Carter, Investigating the antecedents and benefits of SOA implementation: A multi-case study approach, AMCIS Proceedings, Paper 195, 2007
10. S. Aier, T. Bucher, R. Winter, Critical success factors of service orientation in information systems engineering. Bus. Inf. Syst. Eng. **3**, 77–88 (2011)

# Chapter 6
# Conclusion and Service-Oriented Architecture (SOA) Experts Recommendations for Organizations

## 6.1 Introduction

This chapter emphasizes on the discussion, conclusion, and recommendations of this study. The main goal of this study was to identify the significant factors effects on SOA adoption in organizations and to clarify the impact of SOA adoption on the performance of organizations. Besides, this study proposed a suitable framework for adoption of SOA in organizations which seems to be beneficial for future scholars. The following points are discussed in this chapter:

- Discussion of Findings
- Achievements
- Contribution
- Recommendation
- Limitation and Future Work
- Chapter Summary

## 6.2 Discussion of Findings

As it mentioned several times during this study, the main aim of this study is to find the significant factors which are affected on SOA adoption in organizations and to present an SOA adoption framework to show the relationship between key factors and SOA adoption and the performance of organizations. Eight factors are extracted from previously published papers from 2009 till 2013. These factors are: (1) complexity, (2) security, (3) strategy, (4) culture and communication, (5) governance, (6) business and IT alignment, (7) return on investment, and (8) costs.

In order to investigate the significance of these factors in organizations, author proposed a framework which focuses on the relationship of these factors to increase the opportunity of being successful in SOA adoption. For data collection self-selected sampling is chosen. Through online survey questionnaires are distributed among SOA professionals. The result of this survey shows that governance, strategy, culture and communication, business and IT alignment, and ROI have positively influence on SOA adoption while complexity, security concerns, and costs negatively affected SOA adoption. The consequences of these analyses are in line with previous researches which are concentrated on the key factors which affected SOA adoption.

## 6.3   Achievements

The researcher tried to achieve and direct the objectives of this study as following list:

- To identify the factors influenced by adoption of SOA in organizations.
- To understand the relationship among significant factors, SOA adoption and the performance of organizations.
- To develop recommendation towards success of SOA adoption.

### 6.3.1   Achievement of Objective 1

**To achieve the first objective of this research**   For gathering data and information, a vast number of papers related to SOA adoption and factors effect on SOA adoption were reviewed. Reviewed papers are found out from various sources like journals, books, etc. This study may help those organizations which are not adopted SOA yet, to find out which aspects of their organization will increase. Researcher endeavored to extract significant and the most potential factors in adopting SOA. Therefore, the author selected eight factors which have the most number of repetitions in previous studies. Most researchers itemized these factors as critical success factors.

### 6.3.2   Achievement of Objective 2

**To accomplish the second objective of this study**   Based on the data and information collected from literature review, a framework was proposed. Before distributing the questionnaire, hypotheses were discussed. And a pilot study was conducted. Questionnaire was distributed among SOA experts through LinkedIn. The reliability and validity of the questionnaire were tested to prepare it for the main data collection and analysis. Finally, questionnaire was distributed among SOA experts based on the proposed framework. The relationship between key factors and SOA adoption and their effect on the performance of organizations were analyzed. For validation and finalization of the proposed framework, researcher used SmartPLS software. The consequences of analyses showed that all hypotheses were accepted. It is concluded that some factors positively effect on SOA adoption like governance, strategy, culture and communication, ROI, and business and IT alignment while others negatively affect SOA adoption such as complexity, security concerns, and costs.

### 6.3.3 Achievement of Objective 3

**In order to accomplish the third objective of this study** Based on the data and information obtained by analyzing the questionnaire and extracted from literature review, some recommendations were provided to guide organizations how to be successful in SOA adoption process. Furthermore, some data were concluded from the experts' experience which filled the last option of the final question in the questionnaire.

The main findings of this research are listed as follows:

- Recognizing the concept, explanation, and the benefits of SOA.
- Identifying the impact of SOA adoption in the organizations.
- Determining and emphasizing the barriers and advantages while adopting SOA in organizations.
- Presenting a suitable framework for adoption of SOA in organizations.
- Highlighting the significant factors that have positive and negative influences on adoption of SOA in organizations.
- Validating the proposed framework of the study.

## 6.4 Contribution

Some researcher emphasized the necessity of a conceptual framework which may assist to investigate SOA adoption and its challenges and barriers. Therefore, this study provided a conceptual framework to show the relationship between key factors, SOA adoption and the impact of SOA adoption on the performance of organizations. The next noticeable contribution is that this study extracts most significant and critical success factors from previous studies since 2009 till 2013. This may help organizations to know the most important factors that influence SOA adoption on their organizations.

Besides, the findings of this study will have some advantages for those companies that have been using SOA to improve their strategies by reviewing the recommendations of SOA experts which are provided in the next section. Moreover, it is beneficial for those companies which do not change their traditional systems, since the results of analyzing data show that more than half of SOA experts were believed their SOA adoption was partially successful while a third of them agreed their SOA adoption is completely successful in organization.

## 6.5   Recommendation

As it is expressed before, the third objective of this study was to develop some recommendations towards success of SOA adoption. For achieving this purpose, a question was adopted in the questionnaire to ask SOA professionals about their recommendations towards success of SOA adoption. According to all recommendations resulted in this study, the following recommendations could be concluded:

- Starting with a small scope, growing up SOA adoption with small steps, and increasing gradually
- Training SOA to all IT levels, creating an organizational culture for supporting SOA
- Establishing an SOA roadmap at the beginning of the adoption
- Using top-down and driven from IT strategy approach
- Communicating between business and IT departments besides aligning

Moreover, three SOA experts mailed their recommendations and ideas directly through LinkedIn to the researcher. As these ideas were useful and helped the researcher to think deeply about the success of SOA adoption, so they are listed as below:

1. A top manager of consulting services company stated that:

- "When it comes to the cost of projects SOA leads to increased costs as service development is more complex than other development. In time these costs might be gained back by things as agility and maintainability (due to a clear set of rules and well-known services) and perhaps a little reuse."

2. An IT manager in telecommunication industry expressed that:

- "It is important to know that SOA is very prominent to companies but does not mean cost reductions as it promises. SOA has a lot of hide challenges to face such as configuration control, environment availability, development dependency, and etcetera. In my opinion, SOA helps the business performance by providing flexibility and easy adaptation. SOA leads to business agility. But it is very important to know the value chain of each service to know the real revenue. This is very difficult if IT do not work in strong relation with business people."

3. A Chief Executive Officer who wrote a book about SOA declared that:

- "10 years after my book I feel more or less disappointed of the progress SOA made. Still believing that the concept of a service is right it is evident that we are lacking important ingredients for successful SOA adoption. Today I believe that ethics and corporate culture are the key areas we have to look at. SOA is only feasible if your company is service-oriented—i.e. it truly strives for customer value."

In addition, since measuring ROI is time consuming with some difficulties, so it is recommended to deploy SOA in a small scope of organization and estimating ROI before going to the enterprise wide.

## 6.6   Limitation and Future Work

As it is mentioned before, this study used non-probability sampling method and self-selecting technique, so the results of this research do not support the general population of worldwide. Anyway, the researcher believed that the sample is representative of organizations which are in the process of adopting or have adopted SOA. The consequences of this study open good occasions for future research in the area of SOA adoption.

- Mixing qualitative and quantitative research method to recognize the correlation between other dimensions of SOA adoption with the key factors which this study focused on.
- A comparative research methodology would also be helpful to compare the performance of organizations which were using SOA and the others which had not adopting SOA.
- The proposed framework would give the idea of measuring the organizational performance more deeply to researchers.

## 6.7   Chapter Summary

According to the objectives of this study, the researcher has illustrated definition and concept of service-oriented architecture adoption. Many case studies and previous studies were surveyed, and significant and potential factors were extracted from literature reviewed. An adopted questionnaire was created, and self-selected sampling method was used for collecting data from SOA experts through LinkedIn.

As a consequence, by analyzing the findings of this study it is proved that all objectives of this study were achieved and the significant factors that effect on SOA adoption were extracted. In addition, the relationship among SOA adoption, key factors and the performance of organization were examined. It is concluded that factors of this study affected on SOA adoption 86% while SOA adoption influences the performance of organization 54%. The main point of this research that can be verified is the recommendations of SOA experts towards success of SOA adoption in organizations. This may help organizations to follow professionals' suggestions during their migration from traditional systems to implementing SOA.

# Appendix A: Online Questionnaire

- ⊙ Construction/Engineering
- ⊙ Distributor
- ⊙ Education
- ⊙ Electronics
- ⊙ E-marketplace (portals, auction, vert.)
- ⊙ Energy
- ⊙ Financial services/Banking
- ⊙ Food/Beverage
- ⊙ Government
- ⊙ Healthcare/Medical
- ⊙ Hospitality/Travel
- ⊙ Insurance/HMO
- ⊙ IT Vendor
- ⊙ Logistics/Transportation
- ⊙ Manufacturing/Industrial (non-computer)
- ⊙ Media/Entertainment
- ⊙ Metals & Natural Resources
- ⊙ Non-profit
- ⊙ Retail/E-commerce
- ⊙ Telecommunications/ISP
- ⊙ Utilities
- ⊙ Other: [＿＿＿＿＿＿＿＿＿]

**3. Please estimate how many employees in total are in your organization? \***

- ⊙ Less than 50
- ⊙ 50 - 99
- ⊙ 100 - 499
- ⊙ 500 - 999
- ⊙ 1,000 - 4,999
- ⊙ 5,000 - 9,999
- ⊙ 10,000 or more

N. Niknejad et al., *The Impact of Service Oriented Architecture Adoption on Organizations*, SpringerBriefs in Electrical and Computer Engineering, https://doi.org/10.1007/978-3-030-12100-6

**4. Overall, how would you describe your success or failure of SOA projects in your organization?**

◎ Successful: We have completed or will soon complete the project as planned and achieve(d) all the desired goals

◎ Partially Successful: We have completed or will soon complete the project and achieve(d) most of the goals

◎ Not Successful: We did not complete the project and did not achieve the desired goals

◎ A Fiasco: The project caused significant disruption, with no benefit

◎ It's too early to tell

**5. Which of the following best describes your organization use of, service-oriented architecture (SOA)? ***

◎ Not pursuing and no immediate plans to do so

◎ Planning to pursue SOA within the next 6 months

◎ SOA projects are in the pilot stage

◎ SOA projects are in development

◎ Deployed in production for use in a single department

◎ Deployed in production for use in multiple departments

◎ Deployed in production at enterprise level

**6. What approach have your organization used to starting SOA initiative? ***

◎ Top-down from business strategy to IT strategy

◎ Bottom-up from IT strategy to business strategy

◎ Driven from IT (architecture) strategy

◎ Other: [                    ]

[ Continue » ]                          [                              ]

33% completed

# Section 2 - Technology Context

The following questions ask you about the technological aspects of your SOA implementations.

### a. Complexity *

SD=Strongly Disagree; D=Disagree; N=Neutral; A=Agree; SA=Strongly Agree

|  | SD | D | N | A | SA |
|---|---|---|---|---|---|
| 7. Learning to use SOA technologies and associated standards is complicated. | ◉ | ○ | ○ | ○ | ○ |
| 8. Choosing the right standard for our SOA implementations required lots of researching and prototyping. | ◉ | ○ | ○ | ○ | ○ |
| 9. In general SOA is very complex to use. | ◉ | ○ | ○ | ○ | ○ |

**b. Security \***

(S D=Strongly Disagree; D=Disagree; N=Neutral; A=Agree; S A=Strongly Agree)

|  | S D | D | N | A | S A |
|---|---|---|---|---|---|
| 10. SOA implementations within our organization are supported with secure infrastructure. | ◎ | ◎ | ◎ | ◎ | ◎ |
| 11. SOA provides secure services in our organization. | ◎ | ◎ | ◎ | ◎ | ◎ |
| 12. Personally, I do not have any concern about the security and privacy of implementing SOA. | ◎ | ◎ | ◎ | ◎ | ◎ |

**c. Costs \***

(SD=Strongly Disagree; D=Disagree; N=Neutral; A=Agree; SA=Strongly Agree)

|  | S D | D | N | A | S A |
|---|---|---|---|---|---|
| 13. SOA decreases the investment in new IT project. | ◎ | ◎ | ◎ | ◎ | ◎ |
| 14. SOA eliminates the cost of | | | | | |

upgrading the
legacy system.

○      ○      ○      ○      ○

15. SOA
decreases the
cost of system
maintenance.

○      ○      ○      ○      ○

« Back    Continue »

66% completed

# Section 3 - Organization Context

The Following questions ask you about your organizational aspect of your SOA
implementation.

**a. Culture and Communication \***

(SD=Strongly Disagree; D=Disagree; N=Neutral; A=Agree; SA=Strongly Agree)

|  | S D | D | N | A | S A |
|---|---|---|---|---|---|
| 16. Paying attention to establish communication between business and IT department is very important. | ○ | ○ | ○ | ○ | ○ |
| 17. Our employees willing for change towards SOA adoption. | ○ | ○ | ○ | ○ | ○ |
| 18. Adequate communication of all service orientation |  |  |  |  |  |

project-
stakeholders
effect on SOA
adoption.

| | | | | |
|---|---|---|---|---|
| ◎ | ◎ | ◎ | ◎ | ◎ |

**b. Governance \***

(SD=Strongly Disagree; D=Disagree; N=Neutral; A=Agree; SA=Strongly Agree)

| | S D | D | N | A | S A |
|---|---|---|---|---|---|
| 19. Our organization has established SOA governance that is fully integrated within IT governance. | ◎ | ◎ | ◎ | ◎ | ◎ |
| 20. Definition of organizational responsibilities for managing the service landscape is critical for our organization. | ◎ | ◎ | ◎ | ◎ | ◎ |
| 21. Definition of processes for service development and service adaptation is very important | ◎ | ◎ | ◎ | ◎ | ◎ |

in our
organization.

22. Definition
of service
ownerships is            ○          ○          ○          ○          ○
vital in SOA
adoption.

**c. Strategy** *

(SD=Strongly Disagree; D=Disagree; N=Neutral; A=Agree; SA=Strongly Agree)

|  | S D | D | N | A | S A |
|---|---|---|---|---|---|
| 23. Our organization's IT strategy support application integration with internal and/or external application services. | ○ | ○ | ○ | ○ | ○ |
| 24. Our organization's SOA strategy is dependent on business strategy | ○ | ○ | ○ | ○ | ○ |

**d. Retern on Investment** *

(SD=Strongly Disagree; D=Disagree; N=Neutral; A=Agree; SA=Strongly Agree)

|  | S D | D | N | A | S A |
|---|---|---|---|---|---|

| | S D | D | N | A | S A |
|---|---|---|---|---|---|
| 25. Implementing SOA increased revenue and reduced costs in our organization. | ○ | ○ | ○ | ○ | ○ |
| 26. Overall, SOA Implementation positively increased Return on Investment. | ○ | ○ | ○ | ○ | ○ |

### e. Business & IT Alignment *

(SD=Strongly Disagree; D=Disagree; N=Neutral; A=Agree; SA=Strongly Agree)

| | S D | D | N | A | S A |
|---|---|---|---|---|---|
| 27. The IT strategy is accurately aligned with the business strategy in our organization. | ○ | ○ | ○ | ○ | ○ |
| 28. The IT investments are accurately aligned with the business objectives in our organization. | ○ | ○ | ○ | ○ | ○ |

| 29. The business strategy is effectively supported by the IT strategy in our organization. | ○ | ○ | ○ | ○ | ○ |
|---|---|---|---|---|---|

# Section 5. SOA Adoption & Organizational Performance

The following questions ask you about the effect of SOA adoption on the performance of your organization.

### a. SOA Adoption *

(SD=Strongly Disagree; D=Disagree; N=Neutral; A=Agree; SA=Strongly Agree)

|  | S D | D | N | A | S A |
|---|---|---|---|---|---|
| 30. Moving to a Service Oriented Architecture (SOA) brings many benefits to businesses. | ○ | ○ | ○ | ○ | ○ |
| 31. Agility for business is delivered by aligning IT infrastructure with business requirements through a well implemented | ○ | ○ | ○ | ○ | ○ |

SOA.

**b. Organizational Performance ***

(SD=Strongly Disagree; D=Disagree; N=Neutral; A=Agree; SA=Strongly Agree)

|  | S D | D | N | A | S A |
|---|---|---|---|---|---|
| 32. Implementing SOA positively increases the performance of our organization. | ○ | ○ | ○ | ○ | ○ |
| 33. Adoption of SOA leads our organization to better performance. | ○ | ○ | ○ | ○ | ○ |

# Recommendation

**34. What is your recommendation to be successful in SOA adoption process based on your own experience? ***

(Multiple Choices)

☐ Develop an architecture with a vision for the future.

☐ Foresee linkages from IT to your business processes.

☐ Create an organizational culture and skills to support SOA.

☐ Build a scalable infrastructure.

☐ Enable operational visibility through governance and service management

☐ Other: [                    ]

**35. If it is possible for you please cite your company name just for validating your answers for my examiners.**

```
┌──────────────────────────┐
│                          │
└──────────────────────────┘
```

**36. Please specify the name of your country. ***

```
┌──────────────────────────┐
│                          │
└──────────────────────────┘
```

[ « Back ]  [ Submit ]

Never submit passwords through Google Forms.                    100%: You made it.

# Appendix B: Factors Influence on SOA Adoption

N. Niknejad et al., *The Impact of Service Oriented Architecture Adoption on Organizations*, SpringerBriefs in Electrical and Computer Engineering, https://doi.org/10.1007/978-3-030-12100-6

| | Factor | 1 Lawler et al. | 2 Luthria and Rabhi | 3 Ciganek et al. | 4 Galinium and Shahbaz | 5 Antikainen and Pekkola | 6 Chang and Lue | 7 Vegter | 8 Lee et al. | 9 Joachim et al. | 10 Aier et al. | 11 Caimei Hu | 12 Findikoglu | 13 Seth et al. | 14 MacLennan and Van Belle | 15 Basias et al. | 16 Koumaditis et al. | 17 Emadi and Hanza | 18 Choi et al. | Factory frequency |
|---|---|---|---|---|---|---|---|---|---|---|---|---|---|---|---|---|---|---|---|---|
| | Year | 2009 | 2009 | | | | | | 2010 | 2010 | 2011 | 2011 | 2011 | 2012 | 2012 | 2013 | 2013 | 2013 | | |
| 1 | Awareness | | | | | | | | √ | | | | | | | | | | | 1 |
| 2 | Architecture of the legacy systems | | | | √ | | | | | | | | | | | | | | | 1 |
| 3 | Assessing performance of service processes | | | | | | | | √ | | | | | | | | | | | 1 |
| 4 | Building an industry-wide foundation for SOA | | | | | | | | √ | | | | | | | | | | | 1 |
| 5 | Business and IT alignment | | | | | | √ | √ | √ | | | | √ | | | √ | √ | | | 6 |
| 6 | Business/IT agility | √ | | | | | | | | | | | | √ | | √ | | | √ | 4 |
| 7 | Business driven development | | | | | √ | | | | | | | | | | | | | | 1 |
| 8 | Business partner/customer demand | √ | | √ | | | | | | | | | | | | | | | | 2 |
| 9 | Business process | | | | √ | | | | √ | | | | | | | | | √ | | 3 |
| 10 | Business process management | √ | | | | | | | √ | | | | | | | | | | | 2 |
| 11 | Business stakeholder participation | | | | | √ | | | | | | √ | | | | | | | | 2 |
| 12 | Centralization and formalization | | | | | | | | | | | | √ | | | | | | | 1 |
| 13 | Challenges in scope understanding | | | | | | | | | | | | | √ | | | | | | 1 |
| 14 | Change management | | | | | | | | | | | | | √ | | | | | | 1 |
| 15 | Characteristics of integration projects | | | | | | | | | | √ | | | | | | | | | 1 |
| 16 | Clear goal-setting | | | | | | | | √ | | | | | | | √ | √ | | | 3 |

| # | Factor | | | | | | | | | | | | Count |
|---|--------|---|---|---|---|---|---|---|---|---|---|---|-------|
| 18 | Common language | | | | | | | √ | | | | | 1 |
| 19 | Communication | | | | | √ | | √ | | √ | | √ | 5 |
| 20 | Compatibility | | | | √ | | | | | | | | 2 |
| 21 | Competitive issues | √ | | | | √ | | | | | | √ | 4 |
| 22 | Complexity of SOA technologies | | | √ | √ | √ | √ | | √ | √ | | √ | 6 |
| 23 | Costs | | | | √ | √ | | √ | √ | √ | | √ | 5 |
| 24 | Definition of SOA-based development methodology | | | | √ | | | | | | | | 1 |
| 25 | Dependence on commercial products | | √ | | | | | | | | | | 1 |
| 26 | Deployability | | | | | √ | | | | | | | 1 |
| 27 | Development tools | √ | | | | | | | | | | | 2 |
| 28 | Education and training | √ | | | √ | √ | √ | | | √ | | | 4 |
| 29 | Efficiency and flexibility benefits | √ | | | √ | | | | | | √ | | 3 |
| 30 | Enforce decision | √ | | | | | | | √ | | | | 1 |
| 31 | Enterprise architecture framework | | | | | | | | √ | | | | 1 |
| 32 | Establishing a service development/operation management process | √ | | | √ | | | | | | | | 1 |
| 33 | Executive technology leadership | √ | | | | | | | | | | | 1 |
| 34 | Experience | | | | | √ | | | √ | √ | | | 3 |
| 35 | External process domain on projects | √ | | | | | | | | | | | 1 |
| 36 | External SOA domain on projects | √ | | | | | | | | | √ | | 2 |
| 37 | Fatigue (related to time and workload) | | | | | | | √ | | | | | 1 |
| 38 | Federation | | | √ | | | | | | | | | 1 |
| 39 | Financial benefits | √ | | | | √ | | √ | | | | | 2 |
| 40 | Framing an organizational model for SOA management | | | | | √ | | | | | | | 1 |
| 41 | Governance | √ | √ | √ | √ | √ | √ | √ | | | √ | √ | 12 |

(continued)

**Author legend (column numbers):**

| No. | Author | Year |
|---|---|---|
| 1 | Lawler et al. | 2009 |
| 2 | Luthria and Rabhi | 2009 |
| 3 | Ciganek et al. | 2009 |
| 4 | Galinium and Shahbaz | 2009 |
| 5 | Antikainen and Pekkola | 2009 |
| 6 | Chang and Lué | 2009 |
| 7 | Vegter | 2009 |
| 8 | Lee et al. | 2010 |
| 9 | Joachim et al. | 2010 |
| 10 | Aier et al. | 2011 |
| 11 | Caimei Hu | 2011 |
| 12 | Findikoglu | 2011 |
| 13 | Seth et al. | 2012 |
| 14 | MacLennan and Van Belle | 2012 |
| 15 | Basias et al. | 2013 |
| 16 | Koumaditis et al. | 2013 |
| 17 | Emadi and Hanza | 2013 |
| 18 | Choi et al. | 2013 |

| Factor | Factor | 1 | 2 | 3 | 4 | 5 | 6 | 7 | 8 | 9 | 10 | 11 | 12 | 13 | 14 | 15 | 16 | 17 | 18 | Factory frequency |
|---|---|---|---|---|---|---|---|---|---|---|---|---|---|---|---|---|---|---|---|---|
| 42 | Human Resource | | | | ✓ | | | | ✓ | | | ✓ | ✓ | | | | | | | 4 |
| 43 | Industry concentration | | | | | | | | | | | ✓ | | | | | | | | 1 |
| 44 | Industry fragmentation and inertia | | | ✓ | | | | | | | | ✓ | | | | | | | | 2 |
| 45 | Industry leadership | | | ✓ | | | | | | | | | | | | | | | | 1 |
| 46 | Information architecture | | | | ✓ | | | | | | | | | | | | | | | 1 |
| 47 | Integration architecture and design | | | | | | | | | | | | | ✓ | | | | | | 1 |
| 48 | Integration business and IT | | | | | | | | | | ✓ | | | | | | | | | 1 |
| 49 | Integration strategy | | | | | | | | | | ✓ | | | | | | | | | 1 |
| 50 | Interconnectedness | | | | | | | | | | | | ✓ | | | | | | | 1 |
| 51 | Interoperability | | | | | | | ✓ | | | | | | | | | | | | 1 |
| 52 | IT infrastructure | ✓ | | | | | | | ✓ | | | | | | | ✓ | | | ✓ | 4 |
| 53 | Knowledge exchange | | | | | | | | | | | | | | | | | | ✓ | 1 |
| 54 | Legacy systems integration/architecture | | | | | | | | | | | | | ✓ | | | | | | 1 |
| 55 | Management fad | | | | | | | | | ✓ | | | | | | | | | | 1 |
| 56 | Management | | | | | | | | | | | | | | | ✓ | | ✓ | | 2 |
| 57 | Managing SOA policy processes | | | | | | | | ✓ | | | | | | | | | | | 1 |

| # | Factor | Count |
|---|--------|-------|
| 58 | Market and regulatory differentials | 1 |
| 59 | Measurement | 1 |
| 60 | Migration factors | 1 |
| 61 | Momentum resources and strategic importance | 1 |
| 62 | Multiple standards and platforms | 1 |
| 63 | Naming conventions | 1 |
| 64 | Organization philosophy | 1 |
| 65 | Organizational culture | 6 |
| 66 | Organizational agility | 2 |
| 67 | Organizational management | 1 |
| 68 | Organizational maturity | 1 |
| 69 | Organizational size and scale | 3 |
| 70 | Organizational strategy (long-term planning) | 8 |
| 71 | Organizational structure | 1 |
| 72 | Partner readiness | 1 |
| 73 | Perceived value | 1 |
| 74 | Performance of services-based applications | 1 |
| 75 | Potential implementation challenges | 1 |
| 76 | Potential of the legacy applications | 1 |
| 77 | Process automation | 1 |
| 78 | Processes and methodologies | 2 |
| 79 | Project identification | 1 |
| 80 | Project management | 1 |
| 81 | Procurement/newness of technology | 2 |
| 82 | Putting in place of enterprise-wide architecture management system | 1 |
| 83 | Relative advantage | 1 |

(continued)

| Year | | 2009 | | | | 2010 | | | | | 2011 | | 2012 | | | 2013 | | | | |
|---|---|---|---|---|---|---|---|---|---|---|---|---|---|---|---|---|---|---|---|---|
| | Author | Lawler et al. | Luthria and Rabhi | Ciganek et al. | Galinium and Shahbaz | Antikainen and Pekkola | Chang and Lue | Vegter | Lee et al. | Joachim et al. | Aier et al. | Caimei Hu | Findikoglu | Seth et al. | MacLennan and Van Belle | Basias et al. | Koumaditis et al. | Emadi and Hanza | Choi et al. | **Factory frequency** |
| **Factor** | | **1** | **2** | **3** | **4** | **5** | **6** | **7** | **8** | **9** | **10** | **11** | **12** | **13** | **14** | **15** | **16** | **17** | **18** | |
| 84 | Reliability | | | | | | | | | | | | √ | | | | | | | 1 |
| 85 | Regulatory influence | | | | | | | | | | | | √ | | | | | | | 1 |
| 86 | Resistance to change | | | | | | | | | | | | | | | √ | | | | 1 |
| 87 | Resources sufficiency/competences | | | | | | √ | | | | | | | √ | | √ | √ | | | 4 |
| 88 | Return on Investment (ROI) | | | | | | | √ | | | | | √ | √ | √ | √ | | | | 5 |
| 89 | Reusability of services | | | | | | | √ | | | | | | | | | | | | 1 |
| 90 | Risk management | | | | | | | | | | | | | √ | | √ | | | | 2 |
| 91 | Roadmap | | | | | | | | | | | | | | | | √ | | | 1 |
| 92 | Roles | | | | | | | | | | | | | | | | √ | | | 1 |
| 93 | Security issues (risk, concern....) | | | √ | | | | | √ | | | | √ | √ | | √ | | | | 5 |
| 94 | SOA perceptions | | | | | | | | √ | | | | | | | | | | | 1 |
| 95 | SOA registry | | | | | | | | √ | | | | | | | | | | | 1 |
| 96 | SOA team | | | | | √ | | | √ | | | | | | | | √ | | | 3 |
| 97 | SOA technology/standards | | | | | | | | | | | | | | | | | √ | | 1 |
| 98 | Standards immaturity | | | | | | | | | | | √ | | | | | √ | | | 2 |

| # | Factor | Count |
|---|--------|-------|
| 99 | Step by step evolution planning with consideration of current capacity | 1 |
| 100 | Strengthening business service oriented design process | 1 |
| 101 | Technical skills/expertise | 4 |
| 102 | Technological knowledge | 1 |
| 103 | Technology capability of organization | 2 |
| 104 | Technology planning | 1 |
| 105 | Testing | 2 |
| 106 | Top management support | 4 |
| 107 | Transparency of design artifacts | 1 |
| 108 | Trust between business units | 1 |
| 109 | Trust in web services | 1 |
| 110 | User involvement and organizational commitment | 1 |
| 111 | Vendor support for SOA | 2 |
| 112 | Vendor diversification options | 1 |
| 113 | Web service/XML standards | 3 |
| 114 | Web services best practices | 1 |
| 115 | Stress (related to time, workload, and new requirements) | 1 |

# Index

Printed in the United States
By Bookmasters